Systematic Design of Distributed Industrial Manufacturing Control Systems

Dissertation

zur Erlangung des akademischen Grades

Doktor-Ingenieur (Dr.-Ing.)

Genehmigt durch das

Zentrum für Ingenieurwissenschaften

der Martin-Luther-Universität Halle-Wittenberg

als organisatorische Grundeinheit für Forschung und Lehre im Range einer Fakultät

von

Herrn Dipl.-Ing. Martin Hirsch

geboren am 22.01.1980 in Halle (Saale)

Geschäftsführender Direktor (Dekan): Prof. Dr.-Ing. habil. Dr. h.c. Holm Altenbach

Gutachter:

1. Prof. Dr.-Ing. Hans-Michael Hanisch

2. Prof. Dr.-Ing. Georg Frey

Halle (Saale), 20. August 2010

Reihe: Hallenser Schriften zur Automatisierungstechnik
herausgegeben von:
Prof. Dr. Hans-Michael Hanisch
Lehrstuhl für Automatisierungstechnik
Martin-Luther-Universität Halle-Wittenberg
Kurt-Mothes-Str. 1
06120 Halle/Saale

email: Hans-Michael.Hanisch@iw.uni-halle.de

Bibliografische Information der Deutschen Nationalbibliothek

Die Deutsche Nationalbibliothek verzeichnet diese Publikation in der
Deutschen Nationalbibliografie; detaillierte bibliografische Daten sind
im Internet über http://dnb.d-nb.de abrufbar.

(Hallenser Schriften zur Automatisierungstechnik; 6)

ISBN 978-3-8325-2607-8

Logos Verlag Berlin GmbH
Comeniushof, Gubener Str. 47,
10243 Berlin
Tel.: +49 030 42 85 10 90
Fax: +49 030 42 85 10 92
INTERNET: http://www.logos-verlag.de

Vorwort

Diese Dissertation entstand im Rahmen meiner Anstellung als wissenschaftlicher Mitarbeiter am Lehrstuhl für Automatisierungstechnik der Martin-Luther-Universität Halle-Wittenberg.

Zuerst möchte ich mich bei meinem Doktorvater Prof. Dr.-Ing. Hans-Michael Hanisch bedanken, der mir durch meine Anstellung die Mittel und Möglichkeiten gegeben hat, diese Arbeit erfolgreich zu Ende zu führen. Hans-Michael Hanisch ist mir in der Zeit in Halle nicht nur ein hervorragender akademischer Mentor gewesen, sondern auch ein verständnisvoller Freund geworden, der nie um ein unterhaltsames Gespräch verzagt war, wonach die Weltachse stets neu geschmiert war.

Weiterhin möchte ich mich herzlich bei meinem zweiten Gutachter dieser Arbeit bedanken, Prof. Dr.-Ing. Georg Frey. Ihn lernte ich zunächst auf dem Papier in Form von Veröffentlichungen kennen, später persönlich auf Konferenzen und Tagungen, spätestens aber bei den regelmäßigen Treffen des GMA-Fachausschusses „Methoden der Steuerungstechnik".

Nicht ungenannt bleiben soll Dr.-Ing. habil. Valeriy Vyatkin, der mir während der Anfertigung meiner Doktorarbeit mit fruchtbaren Gesprächen und Anregungen zur Seite stand. Auch er ist mir, spätestens während meines Aufenthalts in Neuseeland, ein guter Freund geworden.

Meine Kollegen am Lehrstuhl für Automatisierungstechnik boten mir während meiner Zeit in Halle immer gutes Teamwork, allein die gemeinsamen Veröffentlichungen sprechen hier für sich. Auch meinem Studienarbeiter und späteren Diplomanden, Herrn Dipl.-Ing. Andreas Helbich möchte ich an dieser Stelle danken.

Nicht zuletzt möchte ich mich bei meinen Freunden bedanken, die während der Entstehungszeit meiner Doktorarbeit stets ein offenes Ohr hatten und mich nie hängen ließen.

Ganz besonderer Dank gebührt meiner lieben Familie, die mich, nicht nur während meiner Doktorarbeit, in erheblichem Maße unterstützt hat und ohne deren Vertrauen und Zuversicht ich wohl nicht hier angelangt wäre. Vor allem meiner Großmutter Gretel möchte ich an dieser Stelle danken.

Halle, im August 2010.   Martin Hirsch

Content

List of Figures

List of Tables

1   Introduction and Problem Description

Networked controllers on the sensor/ actuator level are very common things today. The goal of engineers is to reduce the costs of the design process as well as to increase the functional safety of the designed controllers.

Usually the behavior of the system is described in textual form in requirements specifications. It is a common problem to amend these specifications by means of semiformal descriptions, which are consistently comprehendible by all potentially participating project partners. By using such semiformal descriptions during the design process, incompleteness of information, conflicts and so on can be identified and corrected in a much better way. The goal is to avoid time- and work-intensive corrections in late stages of projects.

The usage of the new IEC 61499 standard is a wish of many research groups that are dealing with this issue and conceive, that major vendors still are blocking the possible future. The advantages could not be more obvious. Numerous contributions can be found in the research community. On the one hand it is an efficient way for implementing distributed, heterogeneous control platforms. The standard covers the potential of designing software components, which are distributed and reusable to a certain extent and offer interoperability of devices, through they are hardware-independent [13]. It is not anticipated that vendors will change their control environments from today to tomorrow, but a smooth integration into existing automation solutions should be a first step to gain the benefits of the standard, which are mentioned in many chapters of this work as well.

Also, many companies are not getting familiar to formal methods. Even the usage of simple State Machines as a description means is no daily business. This concerns defining uncontrolled formal models of the plant as well as of the controllers. These formal models are a means to have a formal model of the closed-loop behavior, which can be analyzed with means of model-checking, i.e. analysis of reachability. With formal verification methods faults can be detected, which are difficult to detect with pure simulation and difficult to predict in advance. For instance, this could be states, which must be eliminated or avoided because of important safety concerns.

This work tries to meet the issues mentioned above with a framework, whose steps have the potential to be smoothly linked together but all of them can also be used in a single context.

In this work, a complete system engineering framework is presented, regarding the fundamental key words like distribution and complexity, but still reliability. The design of this framework is based on strong symbioses between the single steps during the entire engineering process of a distributed system and is leaned on [1], which has been developed in the work group in Halle.

In contrast to prior work [1], where the similar method was proposed, this work is beyond the phase of just proposing the method. The several parts are ready and linked together in this work. The three steps of the framework and their composition were chosen carefully, bearing in mind current results from research and industry. So, it is possible to adopt the framework by using up-to-date tools. The work should be of interest in the automation industry sector as well as for any other company designing distributed, embedded controllers.

The enhancement of written specifications/ requirements by using well-chosen diagrams, which can be understood immediately by engineers, is just the first step of the framework. This step then allows generating Function Blocks and even entire System Configurations according to the IEC 61499 standard. It is no must, but certainly eases the draft of controllers by using already defined structural and behavioral descriptions according to the specifications. The rounding of the framework is the verification of the controllers in closed loop with well-structured models of the plant.

This work has a methodical nature. Although it is accomplished, the linking between the steps of the engineering framework is just a secondary issue. This work primarily deals with systematic methods applied in each step, bearing in mind all of the mentioned issues above, but also to be able to smoothly link the steps together.

To meet the anticipated character of this work, for each chapter an evaluation of the intermediate results is provided for a better understanding of the framework and its construction.

The usage of UML/ SysML along with the IEC 61499 standard has inspired many contributions. In [69], the authors suggest a profile to integrate Function Blocks into the UML. Therefore, the linking between Function Blocks and classes of the UML is done via Adapter Interface Function Blocks. On the one hand, the Adapter interfaces provide the typical Event- and Data-interface, on the other hand they provide UML-specific ports to be connectible with UML classes. The idea is to enhance common UML tools by this technology to be able to link UML and IEC 61499.

In [25], the authors describe a framework to gain entire IEC 61499 system configurations by specifying the system using typical UML diagrams, like Interaction- or Class diagrams. They further developed an IEC 61499-compliant development tool which allows specifying the system using Use cases, Class diagrams, Sequence diagrams and so on.

In [52], the author describes the automatic transformation of component diagrams to block interfaces and networks of Composite Function Blocks, as well as the transformation of State Machines into Execution Control Charts (abbr. ECC). In [10], the authors describe networks of Function Blocks and resources using Component diagrams. Class diagrams are used to specify the hierarchical structure of devices, resources and Function Blocks. In [38], the authors show a structural description of Function Blocks as well as their behavior by using Activity diagrams.

In [70], the author describes an interface of a Function Block by using a UML-class, which is enhanced by stereotyped attributes for in- and outputs.

In [26], the authors describe a transformation mechanism for Function Blocks using State Machines, Sequence diagrams and Class diagrams. State Machines are used to describe ECCs, Sequence diagrams are used to link Function Blocks, devices and resources. Class diagrams are used to describe the structure of Function Blocks.

Despite of numerous works of the community and lots of interesting results there is no observable *systematic* in designing distributed controllers following the IEC 61499 standard, which can be used again and again. In this work it is therefore concentrated on systematic controller design, which shall be applicable to most manufacturing control systems.

For implementing IEC 61499 compliant applications, a couple of tools have been publically made available during the last decade. Thanks to those tools, the emerging standard became a scope in the research area and, significantly to be mentioned, for students, just because the trial implementations made the entrance of the standard into the teaching area possible. In the following, a short survey of the most common tools that have been developed in the last years is presented.

First of all, one needs to mention the Function Block Development Kit (abbr. FBDK) implementation by J.H. Christensen [33]. That tool is widely used in research trial implementations. A lot of Service Interface Function Block (abbr. SIFB) implementations have been developed, so the tool can be used in combination with lots of different Java-based control devices. A short introduction can be found in [50] and [15]. The CORFU Engineering Support System by the work group of K. Thramboulidis [30] supports an UML-based design approach of IEC 61499 compliant system design. Also to mention is the O³NEIDA workbench [32], which includes the concept of Automation Objects [1]. Last but not least the 4DIAC development environment [34] is listed, developed at the Vienna University of Technology, which is a C-Code based variation. A remarkable industrial, commercial tool implementation is ISaGRAF. This software combines both IEC 61131 and IEC 61499 compliant implementation methods [31].

Application of the IEC 61499 standard is not only confined to the industrial manufacturing sector. Although developed for that field, one also comes across example implementations from automobile comfort electronics sector [16] and aviation electronics sector [51]. But basically the standard has been developed for industrial manufacturing systems, and basically the most benchmarks have been developed on that background, although they have not yet reached it completely. Many work groups have been dealing with IEC 61499 compliant test beds. Due to those activities, naturally a certain amount of benchmarks has been developed and commissioned. Thereby one can differentiate between pure academic test beds, and, on the other hand, rare industrial-like prototypes. At first the academic test beds are specified. The work group from Halle started implementing prototype benchmarks for IEC 61499 compliant controllers in 2003 [50]. Through the years, the methodologies advanced in many directions. Starting with systematic building of models of the plant for a sound visualization, also the controller design methodologies were further developed. Based on simple centralized controllers, systematic methodologies for implementing distributed controllers were developed. Controller design methods have been published in [50], [36], [15] and [43] (chronological). Above all, the issue of reusing controllers in flexible automation unavoidably led to the design of hierarchical control patterns. The current state of the art benchmarks of the work group in Halle can be found on the site of the group [5]. Also the group of G. Frey developed so called functional mechatronic controllers [52], where the design approach is similar to the one in this work. Further, the academic test beds from the University of Auckland and Vienna University of Technology are to be specified. In a few years, the group of V. Vyatkin created an IEC 61499 compliant test bed laboratory [18]. Reconfiguration on the fly, reusability of distributed control components and verification of the control software are major points of interest in the research of this work group that actually originated in the work done in Halle. Furthermore, a prototype-industrial luggage conveyor system case study taken from an airport vendor is to be mentioned [53]. The work group of the Vienna University of Technology possesses a big lab, dealing with distributed control methods, among others the IEC 61499 standard [19]. Furthermore, the group makes a point demonstrating future scenarios for application of IEC 61499 [54]. A rather industrial-like application of the standard can be found in [55]. In contrast to IEC 61131 frameworks, the authors recognized an effective high-level view of the application. Furthermore, the development process showed up to be more rapidly due to reusing and adapting existing Function Blocks.

The work is structured as follows. The next chapter introduces the engineering framework along with a brief testbed description. Chapter 1 describes the specification of the system, and further the basic principles of the IEC 61499 standard will be provided. In Chapter 4, a design pattern and some approaches for designing distributed, highly reusable controllers are introduced. This chapter is also essential for the verification framework described in Chapter 5. Chapter 6 makes some remarks on integration of the framework into existing automation environments, and finally in Chapter 7 some conclusions are drawn, and a forecast on prospect works is done.

2   Framework Description

The Engineering Framework consists of three important steps, schematically illustrated in Figure 1, without explaining used languages or tools. The important step of implementation and execution is not as deeply explained in this work, as hardware development and hardware access is a large field and not the major interest in this context.

Step 1 is to capture the requirements of a system and to embed them into a consistent specification, consisting of several, easily intuitively understandable diagrams. Step 2 is to implement a hardware-independent, yet executable system specification. It is notable, that the other steps are arranged around this step, which shows, that the IEC 61499 standard can be regarded as an epicenter of the framework. Step 3 will contain a verification framework. This is naturally done iteratively with step 1. The details of the steps are specified in the next subchapters.

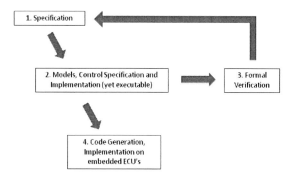

Figure 1: An illustration of the presented engineering framework.

2.1   System Specification

It is no question that the quality of control software strongly depends on the quality of the specification. For the capture of the requirements and the building of a consistent specification, in this work it is for this reason decided to use the Systems Modeling Language (abbr. SysML) [2].

Nowadays, specifications in form of textual descriptions are still widely used. It is not yet commonly used to generate code from specifications or even to do specification simulations, although rich tools are available, as [3], [4].

To keep a specification clearly arranged and to embed important requirements, the SysML offers beneficial diagram types, which enable the change of structures, behavior or just single parameters of the system and its functionality, even in late stages of projects.

The possibility to generate code from the specification according to the IEC 61499 standard or any other control code by considering several modeling guidelines makes the usage of this specification means very interesting for development engineers.

A deeper description of this step and a more precise explanation on the usage of the diagrams will follow in Chapter 1. A complete specification of one station of the running example [5] is described there.

2.2   Control Specification

The implementation of the hardware-independent, yet executable system specification is performed by using the IEC 61499 standard [6]. This emerging standard has been developed on the foundations of the long lasting IEC 61131 standard [7] and shall support such features like portability, reconfigurability, reusability and interoperability. The IEC 61499 standard is especially designed for distributed systems. The architecture represents a light-weight component solution that provides all essential features such as encapsulation of semantics from a particular platform, portability thanks to the unified XML description, reconfiguration under consideration of certain modeling guidelines and a holistic view on distributed applications. Reconfiguration in that context means reconfiguration of the plant architecture or change of the process sequence. There are lots of works dealing with this emerging standard, e.g. [9], [10] and [11] and also the work group of Halle has been dealing with the issue for a long time [12], leading to complete standard works, like [13]. However, just the feature of reusability is a specific point of interest in this work and will be addressed deeply in Chapter 4.4, introducing concepts for designing distributed controllers which end up in hierarchical control concepts.

A brief introduction to the basic principles of this standard will follow in Chapter 3.2.

2.3   Verification

Once the system is modeled, controlled and visualized with IEC 61499 Function Blocks, the next step, formal verification, can be performed. It does not need to be done, but it is just an additional option to increase the quality and safety of system design that should not be underestimated. Any verification method requires a formal model of the closed-loop behavior of the controller and the controlled object. There is a formal model that has been developed in the work group from Halle [14] that is used in this work. This kind of model has been chosen, because it supports encapsulation, modularity, interactions of subcomponents by means of Events and Data and many other features that are relevant in this context. The models are based on (augmented) Petri nets with a signal interface given by Condition and Event inputs/ outputs. They are therefore called Net Condition/ Event systems (abbr. NCES) or Condition/ Events nets (abbr. C/ E nets). They have a precisely defined syntax and semantics and therefore allow formal verification, i.e. the mathematical proof that the system fulfills formal specifications. Furthermore, since IEC 61499 controller designs and NCES are very similar, semi-automatic transformation of IEC 61499 specifications based on the Model-View-Control design pattern (see Chapter 4.1) to formal models of the closed-loop behavior is possible. This includes formal verification into the system design process in a very natural way. In this way, the quality of the whole design process is enhanced significantly. Systematics for a verification framework are introduced in Chapter 5. A short introduction to the basic principles of NCES can be found in this chapter as well.

2.4   Implementation and Execution

For the execution of the implemented, verified controllers, the models of the plant introduced in Chapter 4.2 are substituted with adequate Service Interface Function Blocks as described in [15]. Then, the control Function Blocks are executed on the embedded device. Thanks to the portability of IEC 61499 Function Blocks, the developed controllers can be distributed as desired on numerous, heterogeneous embedded devices [15].

The models as well as the visualization components can be executed either on an appropriate embedded device itself, or on an engineering and simulation station representing either a node in a network [15] or even a portable embedded device, supporting wireless communications to perform error location or maintenance jobs [16]. These possibilities have the potential to enhance predictive control strategies and to make online-monitoring activities easier.

Examples of running distributed, heterogeneous control environments can be found in [5] and [17], as well as in [18] and [19].

2.5   Running Example

The testbed on which the issues of this work are shown is a laboratory-sized plant, which consists of four stations as shown in Figure 2. The stations themselves can be divided into several modules. In general, the purpose of the exemplary testbed is to ensure material flow between the stations and their modules in form of work pieces. Further, the work pieces are sorted and processed. The purposes of the single stations are described as follows.

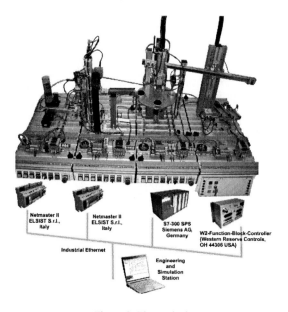

Figure 2: The testbed.

2.5.1   Distribution Station

The purpose of the distribution station (Figure 3) is to provide the subsequent station with work pieces. Therefore, it consists of two mechatronic modules, namely the feeder module with a work piece magazine (1) and the transfer module (2), which transfers the work pieces to the subsequent station. The work piece magazine provides the output *magazine_empty*, in case there are no more work pieces available. The feeder module has the input *extend* followed by the outputs *extended* or *retracted*. Once the feeder has reached the position *extended*, the transfer unit will move to the left position and suck in a work piece with a vacuum valve. Therefore, it has the inputs *to_magazine*, *to_next_st*, *vacuum_on* and *vacuum_off* and the outputs *at_magazine*, *at_next_st* and work piece *wp_sucked_in*. Furthermore, the station is equipped with a panel that includes several buttons to provide operations like reset, start or stop the station. The distribution station is controlled by an IEC 61499 compliant NETMASTER 2 SNAP [20]. The distribution station serves for the specification framework explained in Chapter 1.

Figure 3: The distribution station.

2.5.2   Testing Station

The purpose of the testing station (Figure 4) is to identify the work pieces, to check their height and to sort out incorrect work pieces. What kind of work pieces are submitted to the subsequent station or become sorted out, depends on the specification. To accomplish these issues, the plant offers the following modules. The detection module (1) checks if the work piece is present and black, present and metal or present and red. Therefore, it provides the sensor outputs *wp_present*, *wp_metal* and *wp_not_black*. Once a work piece is identified, the pusher module (2) can directly push the work piece out, according to the specification. Therefore, the pusher module comes with the input *extend_pusher* and the output *pusher_retracted*. Once a work piece is identified as classified, the lifting module (3) transfers the work pieces to the upper position. The station therefore is equipped with the inputs *move_up*, *move_down* and the outputs *lift_up* and *lift_down*. In the upper position, the measuring module (4) checks the height of the work pieces. This module consists of an analogue sensor, which measures the thickness of the work pieces. The functional principle is based on a linear potentiometer with voltage divider tapping. The analogue measured value is processed via the controller with analogue inputs. The measuring module provides the input *eject_cylinder* and the output *cylinder_ejected*. If the measured height fits the defined borders, the work piece is pushed out by the pusher module as described above and moves along the slide module (5), where it is stopped by a stopper until the subsequent station is ready to accept the work piece. This module provides the input *eject_stopper*. If the height does not fit, the lift will move into the down position and the work piece is pushed out there. Also, the station is equipped with a panel that includes several buttons to provide operations like reset, start or stop the station. The testing station is controlled by an IEC 61499 compliant NETMASTER 2 SNAP. The testing station serves for the coordinator control pattern described in Chapter 4.4.4.

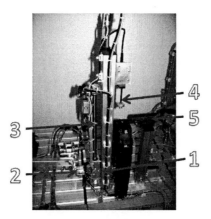

Figure 4: The testing station.

2.5.3   Drilling Station

The purpose of the drilling station (Figure 5) is to drill holes into the work pieces and to check if the hole was drilled correctly. Therefore, it is equipped with several modules. The rotating table (1) ensures that the work pieces are transported between the other modules, the drilling module (2) and the hole checking module (3). First, the work piece arrives from the predecessor station, and the output *work_piece_present* is true. Now the table can rotate and the input *rotate_table* is set to true. Each time the table rotated round 90° the *table_positioned* output is activated and the table shall stop. The work piece is now in the position to be drilled. It is fixed with the negated input *retract_clamp* and once the output *clamp_extended* is true, the drilling machine starts with the inputs *drill_on* and *extend*. Once the output *extended* is true, the drilling machine is moving up with *retract* and then stopped when in the *retracted* position. The clamp is being retracted and if the output *clamp_retracted* is true the work piece is transported to the next, the hole checking module. At this position, the hole checking module is being extended with the negated input *retract_checker*. If the checker is extended (*checker_extended*), the input *retract_checker* is set to true and the module moves into the *checker_retracted* position. Now the rotating table transports the work piece to the transfer position. Furthermore, the station is equipped with a panel that includes several buttons to provide operations like reset, start or stop the station. The drilling station is controlled optionally by an IEC 61131 compliant SIEMENS PLC SIMATIC S7 300 SERIES [21] and alternatively by an IEC 61499 compliant NETMASTER 2 SNAP device. The drilling station serves for the distributed coordinator pattern in Chapter 4.4.5 as well as for the chapter on verification in 5.3.

Figure 5: The drilling station.

2.5.4   Handling Station

The purpose of the handling station (Figure 6) is to take over the work pieces and to handle them either to a slide for work pieces (a) or a slide for work pieces (b). Therefore, it consists of a gripper module (1) and a color-detecting module (2). The gripper module can move to the left and to the right by the inputs *move_left* and *move_right*. It can be lowered respectively lifted with the commands *lower* and *lift_up*. Furthermore, the gripper has the input *open_claw* which is used to grip or release the work pieces. At first the gripper moves left to the predecessor drilling station (*pos_pre_station*). It is lowered (*gripper_lowered*) and checked if a work piece is present (*present*) by an optical sensor. If no work piece is present, the gripper will move up (*gripper_up*). If a work piece is detected, the claw is closed with a negated *open_claw* input. The gripper lifts up and moves to the color-detection module (*pos_detect*) where it is lowered and the detection module gives the ouput *black* in case of a black work piece or *nothing* in case of another work piece with a different color. Then the gripper is lifted and the work piece is submitted to one of the two slides (*pos_slide1, pos_slide2*) and released. The station is also equipped with a panel that includes several buttons to provide operations like reset, start or stop the station. The handling station is controlled by a WRCAKRON W2FBC control unit [22]. The handling station serves for a simple centralized control design as described in Chapter 4.4.2.

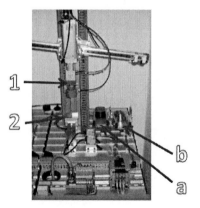

Figure 6: The handling station.

To show the feasibility of integrating the IEC 61499 standard into existing IEC 61131 frameworks, it is decided to control the exemplary testbed with different control units for each station. Three of them are diverse Java-based controllers (to be used compliant to IEC 61499) and one is a common PLC (to be used compliant to IEC 61131). This is to demonstrate the hardware independence and ease of integration of the IEC 61499 standard into existing control environments. No question, the interesting point is the implementation of the needed Service Interface Function Blocks to access the resources of the hardware and the communication between the controllers. This issue is addressed in [15] and will also be described in Chapter 6.

.

3   Specification

3.1   The Systems Modeling Language

The Systems Modeling Language (abbr. SysML) [2] is based on the Unified Modeling Language (abbr. UML) [23] and is designed especially for Systems Engineering. As the UML is very software-oriented and meanwhile provides numerous description means, it is difficult to select the correct diagram type for the particular problem. A reasonable minimization by excluding some description means of the UML, adapting and adjusting diagram types and extending the UML by some other, more important description means, such as requirements and so on, the SysML developed as an independent modeling language which is useful especially for Systems Engineers.

Therefore, in this work the SysML has been chosen for the specification of the system. This description opportunity offers very applicable solutions to arrange requirements, static structures and behavior of a system. The association of all required diagram types in the specification, including requirements, ensures a consistent description of the system and allows validating the specification by simulating the behavior specifications under consideration of, in this context, indispensable diagrams. Of course, the SysML is not only applicable in this specific framework, but also very helpful in other contexts. In some engineering frameworks there is no need to build a complete specification of the system in each direction, it also can be very useful just to support and ease the model-based design of controllers as denoted in [24]. As the tools on the market are of commercial nature and often the SysML is allegorized as just a plug in for common engineering tools, the reader is referred to Chapter 1, where the two richest tools are mentioned.

In this work, however, the building of the system specification has shown up to be straightforward but still powerful enough when divided into four parts as shown in Figure 7 and under application of the listed diagram types. Hierarchical arrangement of the diagrams allows adjusting the detail-level for the beholder, as not every project participant is interested in all technical details. As it is anticipated to automatically generate Function Blocks and other elements of the IEC 61499 standard, certain modeling guidelines have to be considered as explained in the next subchapters, but the general procedure of specification can always be the same, even without considering these guidelines and applying the SysML in other frameworks.

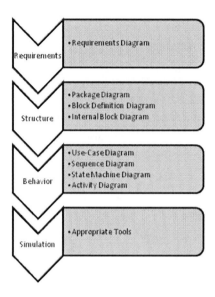

Figure 7: Systematic modeling procedure with SysML.

The gathering of the requirements for a distributed system is a very important step. A requirement depicts one or more characteristics or behavior patterns, which need to be fulfilled at any time. Requirements can be nested, which means that, e.g., one requirement is derived by another requirement or that one requirement enforces another requirement. The more precise the requirements are defined and linked in diagrams, the more detailed the system description can be explained.

The structure of a system results from the inputs and outputs (i.e. sensors and actuators) as well as from offered and required interfaces (communication, energy, etc.). Appropriate diagrams allow a quick survey of the static structure of the system. By unitizing the diagrams it is possible to subdivide a system into several components. The refinement of the structure can then be conducted as necessary.

The description of the behavior of the system is being subdivided into descriptions of the uncontrolled plant and, if desired, the controller itself. The description of the plant specifies the uncontrolled, physical behavior of the plant. The model of the controller controls first all of the models of the plant, such that interaction of the plant and the controller in closed loop is available.

This finally allows the simulation of the complete system for validation, using appropriate tools.

Thus, the static structure descriptions along with the dynamic behavior descriptions are executed and tested against each other. The specification development environment therefore generates executable code, and the developer has the possibility to simulate certain scenarios of the specification. Above all, in bulky and complex projects this brings essential benefits, as already during the phase of the development of the specification, validation can be performed and therefore the specification can be tested on its consistency. The scenarios can be tested either step-by-step or automatically, without intervention of the developing engineer.

Feasible frameworks to converge step 1 with step 2 of the engineering framework are introduced in detail in [25], [26] and [27].

In the following subchapters, an example for a specification following the guideline from Figure 7 is illustrated by means of the example of the Distribution Station described in Chapter 2.5.

3.2   The IEC 61499 Standard

The IEC 61499 standard [6] is especially designed for distributed applications. It defines a reference architecture which includes all essential elements for designing distributed automation systems. The main difference between the IEC 61131 standard [7] and the IEC 61499 standard is the replacement of the cyclic execution model with an Event-based execution model. Function Blocks (abbr. FBs) provide Event- and Data-interfaces and include Event-driven Execution Control Charts (abbr. ECC), which are Event-driven State Machines. It is obvious, that by the mentioned features in Chapter 2.2 the IEC 61499 should support the reusability of controllers. Support of reusability, however, does not mean automatically that any controller that is designed using the IEC 61499 standard is reusable. Engineering methodologies are required, that enable a design of controllers that are reusable to a large extent. This extent clearly depends on the structure and the functionality of the plant that has to be controlled.

The elementary components of the IEC 61499 standard are Function Blocks. As the name announces, Function Blocks encapsulate functionality. A generic example of how Function Blocks are structurally constructed is shown in Figure 8.

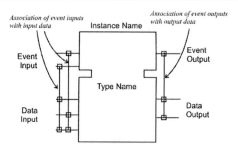

Figure 8: The structure of an IEC 61499 Function Block. (From [13])

The figure shows the generic body of Function Blocks. The head of a FB provides Event-inputs and –outputs. The body of a FB provides Data-inputs and –outputs. Events can be associated with Data as necessary.

There are two types of Function Blocks, *Basic Function Blocks* and *Composite Function Blocks*. Composite Function Blocks can contain numerous other Basic Function Blocks or again numerous Composite Function Blocks. So, it is a component-based architecture.

A Basic FB contains internally an Execution Control Chart (abbr. ECC), algorithms and internal variables as desired. This is schematically shown in Figure 9.

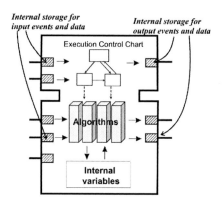

Figure 9: The inside of an IEC 61499 Function Block. (From [13])

The way how a Basic Function Block works is the following. At first, the required input- and/ or internal variables are provided. Then, an input-Event occurs, the associated required variables are made available for the ECC and algorithms are executed as defined. Thereby new variables for the Data-outputs are set and confirmed by an associated output-Event. In case a new Event occurs, the procedure begins again.

An example implementation of an ECC is illustrated in Figure 10. ECCs consist of states and transitions. The states can have assigned algorithms. ECCs contain an initial start state, which is denoted by a double-frame. The transitions between the states can have triggers in form of Events and/ or associated Conditions. If a transition is spontaneous (no trigger or Condition), it is documented with a "1". States can have associated actions in form of algorithms and associated Event-outputs. Algorithms usually process new output Data, and once an algorithm has terminated, this is indicated while leaving the state with an associated output-Event. Events also can be emitted without an associated algorithm. The language that an algorithm is described by in IEC 61499 can be versatile. It contains the known languages of IEC 61131 [8] like Function Block diagram Algorithm, Structured Text, Ladder Logic as well as high level languages such as Java or C.

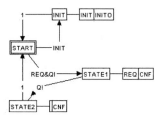

Figure 10: An example of an ECC.

Composite Function Blocks on the contrary do not contain an ECC. They can consist of numerous Function Blocks networked inside, which can be Basic Function Blocks or even Composite Function Blocks as well.

As mentioned above, the IEC 61499 standard further defines a generic model for distributed systems. The standard offers a complete system description as shown in Figure 11. First of all, this includes the controlled process. Then, connected devices and their communication networks are embedded as well as the resources of the devices. Applications can be allocated by engineers as desired without considering device boundaries, at least if needed and feasible. A network of Function Blocks describes an application, which can be allocated by engineers as desired, without bearing in mind borders of devices again. Between Function Blocks there is an Event- and Data flow as indicated in Figure 11 as well.

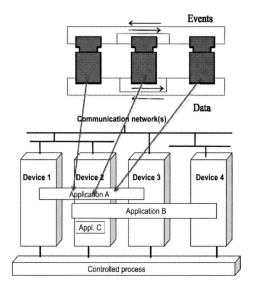

Figure 11: The generic system model in IEC 61499. (From [47])

The connection of the Function Blocks with the devices works by using Service Interface Function Blocks. These Function Blocks allow the access to the resources of the hardware, e.g. to the registers of the I²C Bus of the hardware controller [15]. Service Interface Function Blocks are described briefly in Chapter 6.

The standard led to numerous works that postulate a lot of keywords, which are given in Chapter 2.2 of this work. The following chapters show, how the benefits of the standard potentially can be used and how they should find a way into systems engineering nowadays in a systematic way.

3.3   Requirements

To keep the specification clearly arranged and to hide elements that are not necessarily to be seen at any time, it is useful to allocate the several refining diagrams that will be created during the specification process into packages. Such a Package diagram can be seen in Figure 12. The package [Model] Distribution Station firstly is augmented by the packages system requirements, system structure and system behavior. The system structure is denoted as a package, which in turn has several sub-packages like feed magazine module, transfer module, HMI and ECU. The system requirements and the system behavior are denoted as single packages. Once this frame is specified for the specification, one now can start with the specification of requirements.

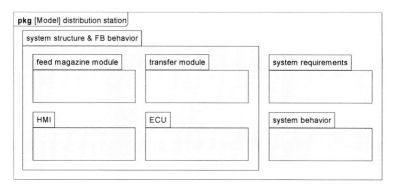

Figure 12: Package diagram of the Distribution Station.

An extension of the UML by the SysML is the possibility to give a description of the requirements by using requirements diagrams. Therefore, requirements and their relations to each other are graphically specified as in Figure 13. The exact content of a requirement is filed in a table, exemplarily shown in Table 1. Requirements can be in certain relations to each other. In this example, only the requirements and sub-requirements as well as the so called derive requirements *(<<deriveRqt>>)* and refine requirements *(<<refine>>)* are beheld. For further kinds of requirement relations it is referred to [28].

In the example shown in Figure 13, the requirements are as follows. The main requirement is the *function of the distribution station*. Its sub-requirements are *supply and separation of work pieces* and *transfer of work pieces*, as well as a Human Machine Interface (*hmi*) and an Execution Control unit (*ecu*). The first and second requirements are now refined in the following. The *checking of the filling level* and *function of the ejecting cylinder* requirements refine the requirement *supply and separation of work pieces*. The requirement *condition of the work pieces* again refines the requirement *checking of the filling level*. The *function of the swivel drive* and *function of the vacuum suction cup* refine the requirement *transfer of work pieces*. As derived relations in the example, the requirements *pneumatic connection* and *coordination of parts* derive from the refine requirements *function of the ejecting cylinder*, *function of the swivel drive* and *function of the vacuum suction cup*. The requirement *power supply* in the example is one with relations to the requirements *hmi* and *ecu*. It could be a sub-requirement, derived from another or refine another requirement, this can be decided as desired as this decision belongs to the developer. In comparison to industrial automation projects, the requirements are fragmentary. These examples are an abstract part of the requirements of this laboratory scaled plant and hence they are not exhaustive, they merely serve for illustration of the used diagram type.

The defined requirements and the according diagram are now inserted into the package system requirements as shown in Figure 12.

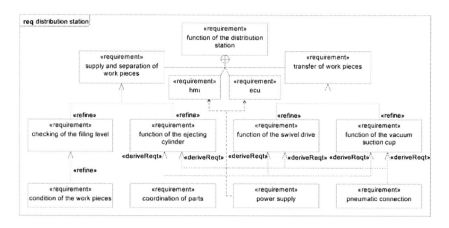

Figure 13: Requirements diagram of the Distribution Station.

Table 1: Description of the requirements.

Name	Text
checking of the filling level	The filling level of the magazine has to be checked by an optical sensor. The feed magazine must be filled manually with work pieces with specified size and material.
Condition of the work pieces	The work pieces consist of plastic or metal material. The plastic ones are either red or black and the metal pieces have a silvery surface. The heights reach from 22,5mm over 25mm to 27,5mm.
coordination of parts	The tasks of the different elements of the station must be coordinated by the controller to prevent clashing of objects. It is not allowed to extend the pusher while the swivel drive is at the position magazine. Dropping the work piece while the swivel drive is still moving is forbidden.
function of the distribution station	The function of the distribution station is to separate one work piece from a magazine and to transfer it to a subsequent station.
supply and separation of work pieces	The station must separate a work piece from others and provide the supply of work pieces for all stations.
transfer of work pieces	A separated work piece has to be transferred to the correct position of the subsequent station.
hmi	The station must provide a Human Machine Interface with START, STOP, RESET buttons.
ecu	For each station exactly one electronic control unit is required.
function of the ejecting cylinder	The double-acting cylinder pushes out the bottom work piece from the gravity feed magazine. Its initial state is the retracted position. It must be able to push a work piece out of the magazine by extending itself and to retract again.
function of the swivel drive	The initial state of the swivel drive is the position magazine. It must be able to reach the work piece, pushed out of the magazine by the ejecting cylinder, and to move to the receive platform of the subsequent station.
function of the vacuum suction cup	The cup must be able to suck in a work piece by generating vacuum and to release the work piece again.
pneumatic connection	The components of the station have to be supplied with compressed air via the service unit. The service unit must be set at 6 bar.
power supply	The station is to be supplied by a 24 V direct current power supply.

3.4 Structure

The structural descriptions shall show the system in a hierarchical way and offer an image of the interfaces of the system, i.e. sensors, actuators, communication interfaces and so on. Therefore, Block diagrams (abbr. bd) and Internal block diagrams (abbr. ibd) are available in the SysML, which derive from Class diagrams respectively composite structure diagrams of the UML. To have the diagrams well arranged, one further has the opportunity to integrate Block diagrams into Package diagrams. In the example, these packages have already been defined in Chapter 3.1.

Blocks can be augmented with Data and Operations. Data means interfaces in form of Attributes (values) according to the physical sensors and actuators, defined by two according Stereotypes. Operations can define Event-inputs/ -outputs or functional actions (algorithms), that in the particular case depends on the Stereotype that the Operation is associated with. The association of Events and the Data-interface is solved with tag-definitions.

Figure 14 schematically shows the hierarchic configuration of the exemplary testbed station. The blocks *Feed magazine module*, *Transfer module*, *Control panel* and *ECU* (Execution Control Unit) are assigned to the block *Distribution station*, thus in an entire-part-hierarchy.

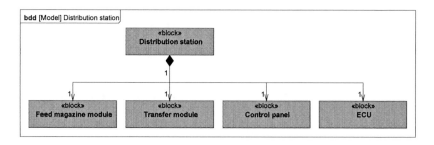

Figure 14: The structure of the Distribution Station.

The level of detailing is not very high so far. To enrich the modules with Data and Operations and more internal refinements, further Block diagrams are created.

It is started with the description of structural interfaces of the exemplary station. As shown in Figure 14, the Distribution station is divided into two mechatronic modules along with a panel and the control unit.

The first module to examine is the feed magazine module. This module consists of two functional components, namely the magazine for providing work pieces and the ejecting cylinder. In Figure 15 one can see the Adapter Interface description of that module. The inputs and outputs are inserted as values. The Operations *REQ* and *CNF* are inserted and associated with the respective inputs and outputs. These Operations later will appear as Event-inputs and –outputs of the Function Blocks in IEC 61499.

The adapter has no functionality or behavior descriptions inside. It only serves for the definition of the inputs and outputs of the module.

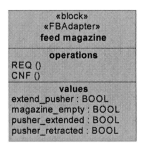

Figure 15: Structural description of the feed magazine module.

Figure 16 shows the adapter description of the transfer module respectively. The Adapter Interfaces defined in this chapter will be connected to the several parts, which are in a close context with the plant itself, namely the models of the plant and the task controllers of the modules (see Chapter 4.1).

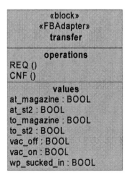

Figure 16: Structural description of the transfer module.

Further, structural descriptions are refined by defining other block definition diagrams. Figure 17 shows the Block diagram of the pneumatic cylinder with magazine module. This diagram contains the blocks for the *ejecting cylinder* and the *gravity feed magazine* with *work pieces*. The single blocks are connected via composite relations. The Operations of the single blocks describe the functionalities of a module (see Chapter 3.5), the Data describe the in- and outputs, respectively the properties of the work pieces. Because it is intended to use these description means for automatic generation of Function Blocks according to IEC 61499, the Blocks are further augmented with Operations like *INIT*, *INITO*, *REQ* and *CNF*. These Operations later will be interpreted as Events according to the IEC 61499 standard.

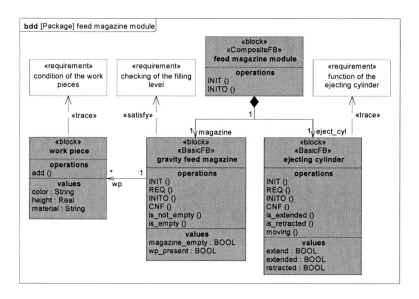

Figure 17: Block diagram of the feed magazine module.

Further worth mentioning are the connections of the blocks with the provided requirements. Connection relations are *<<trace>>* (trace relation) and *<<satisfy>>* (Satisfaction of a requirement).

A fundamental advantage of the association of certain description elements with requirements is the ease of retracing and allocating of the description elements during later on adjustments of the specification. With adequate tools there is the possibility to imply from requirement diagrams to associated description elements, respectively diagrams. Also one can check during simulation (see Chapter 3.6), if every requirement is "touched" and therefore is considered.

The block *feed magazine module* (Figure 18) contains instances of the previously defined adapter *feed magazine* as well as an instance *IN_BOOL*, which just serves for simulation of manually reloading the magazine. The classes *fm1* to *fm8* are interface-classes, which are assigned to Stereotypes *EventConnections* and/ or *DataConnections*, to define Data- and Event-connections between the elements, which will later appear as Function Blocks. A list of the interfaces and their corresponding Stereotypes and the content of their tag definitions can be seen in Table 2.

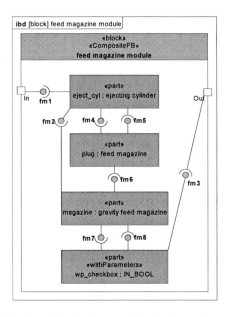

Figure 18: Internal block diagram of the feed magazine module.

Table 2: Overview of the Event-connections of the feed magazine module.

Interface	Event-Connections	Associated Data-Connections
fm1	INIT - eject_cyl.INIT	
fm2	eject_cyl.INITO - magazine.INIT	
fm3	wp_checkbox.INITO - INITO	
fm4	eject_cyl.CNF - plug.CNF	eject_cyl.extended - plug.pusher_extended eject_cyl.retracted – plug.pusher_retracted
fm5	plug.REQ - eject_cyl.REQ	plug.extend_pusher - eject_cyl.extend
fm6	magazine.CNF - plug.CNF	magazine.magazine_empty - plug.magazine_empty
fm7	magazine.INITO - wp_checkbox.INIT	
fm8	wp_checkbox.IND - magazine.REQ	wp_checkbox.OUT - magazine.wp_present

In Figure 19 the detailed Block diagram of the *transfer module* is shown. Here also the blocks are expanded by Data and Operations. Further the blocks are associated again with according requirements.

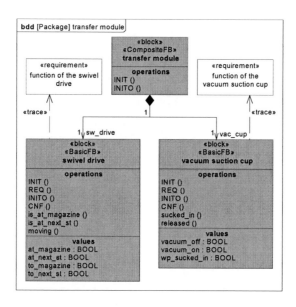

Figure 19: Block diagram of the transfer module.

Figure 20 shows the Internal block diagram of the transfer module respectively.

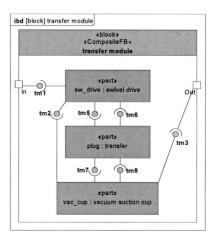

Figure 20: Internal block diagram of the transfer module.

The distribution station further offers a panel and an execution control unit. These devices also need to be structurally described. It is obvious, that the control device has the same inputs as the plant has outputs and the same outputs, as the plant has inputs. Also, the panel outputs are described as inputs for the control device.

As during the phase of specifying the plant it was decided to control it hierarchically in a way described in Chapter 4.4.4, in Figure 21 one can see the anticipated structure of the controllers connected in closed loop with the plant elements.

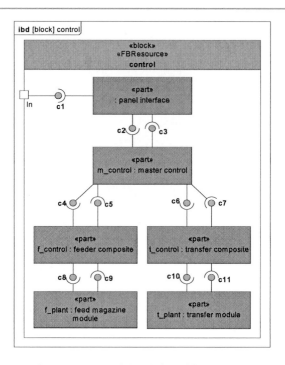

Figure 21: Structural description of the controller.

It is identified in this figure that the controller will be separated into two levels, a master control level and a task control level. These parts are equipped with each bidirectional communication interfaces. The task control parts in turn control the two plant modules of the distribution station separately. That means, the coordinator part coordinates the task controllers, which on the other hand control the functionality of the two modules only. Furthermore, the controller is connected to the panel, which provides the user with the buttons explained in the Use case diagram in Figure 22.

A big advantage of this control architecture is the ability of reuse of the task controllers in each configuration, as only the coordinator needs to be defined again according to the new process specification as it will be specified in the following subchapter.

3.5   Behavior

The behavior descriptions shall help the developer to understand how the automation structure is supposed to work. Based on the structural descriptions defined in the chapter before, the pure static descriptions are now extended by certain behavior descriptions. This contains procedures of desired processes in general, but (can) also contain the desired behavior of single hardware components. An important issue as well is the interaction of the single components.

First of all, the illustration of the Use cases of a system is part of the behavior description. To gain a survey of the expected Use cases, one initially creates a Use case diagram. This contains, beside the user itself, the system including its Use cases (Figure 22), which are illustrated in form of ellipses. In the example of the distribution station, there are the main Use cases *press Button START*, *press Button RESET*, *add work pieces*, *press Button WP_ADDED* and *press Button STOP*, all of them are associated with the operator of the plant. Some Use cases have subsidiarily assigned Use cases associated with the *<<include>>* relation. This relation describes, that one Use case contains another one [28]. In this diagram type, requirements can be associated with Use cases as well.

To describe the Use cases more precisely, one creates the descriptions of the desired system behavior based on the structure descriptions. In this work it is decided to use Sequence diagrams and Activity diagrams for the description of the Use cases. The functionalities of the in Chapter 3.4 defined structure descriptions are later described with State Machines, which are associated directly to the blocks. These descriptions contain the uncontrolled physical behavior of the plant as well as the description of the behavior of the controller.

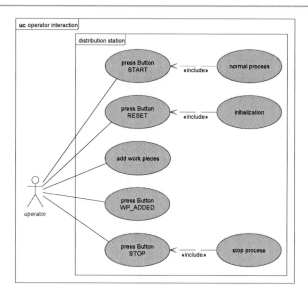

Figure 22: Use case diagram.

By specifying the behavior of a system one describes the expected behavior of the plant to be controlled on the one hand, and on the other hand the behavior of the controller itself, which can be designed either centrally or distributed.

Therewith, the specification differs from formal modeling of a system, where the models represent the uncontrolled physical behavior of the plant.

If one has a look at the Use case diagram, one can see the Use cases, which are directly associated with the user itself, namely pressing the buttons available at the panel. These Use cases include the essential Use cases for the plant, namely initialization, running normally and being stopped.

At first, the Use case *initialization* is specified. First of all, it is a good choice to design Activity diagrams of the Use cases, as they are easy to create and easily understandable. The Activity diagram of the Use case *initialization* is shown in Figure 23.

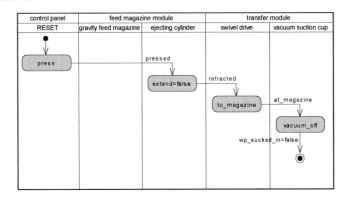

Figure 23: Activity diagram of the Use case initialization.

In this figure, one can see what should happen during the process of initialization. The reader of the diagram starts with the starting node, illustrated as a black point. After pressing the RESET button, the ejecting cylinder has to retract, the vacuum suction cup needs to be turned off and the swivel drive needs to move to the magazine position, all activities in this certain sequence, ending with the ending node, illustrated by a circle with a filled point inside. The transitions between the actions symbolize the sensor values for confirming the completion of an activity. This diagram type is well applicable when it is important to conclude the interaction of the Data values.

A deeper description of this Use case is shown in the Sequence diagram in Figure 24. This diagram shows the principal chain of this Use case. It is anticipated to read this diagram when defining the coordinator control part.

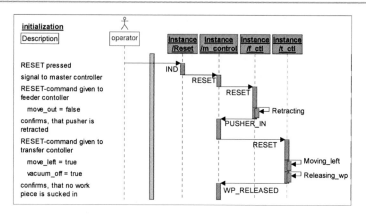

Figure 24: Sequence diagram of the Use case initialization.

This diagram graphically defines the interactions of the control instances, which will be defined precisely later on. The pseudo-algorithms used on the left side of the diagram develop from the particular Activity diagrams for that Use case. It can be seen here, that the operator presses the button *RESET*, and the process of initialization starts. The coordinator control part, here shown as instance *m_control*, makes sure that both of the hardware parts, feeder module and transfer module, move into a position for being ready to start a normal procedure of a process. This is done by the coordinator by sending reset commands first to the feeder controller (*f_ctl*), awaits its confirmation of being reset, and then to the transfer controller (*t_ctl*), also awaiting its confirmation. The difference between the diagrams in Figure 23 and Figure 24 is that the first illustration generally shows the interactions of the actuator values and sensor values (desired behavior of the plant), while the second illustration shows the desired behavior of the control objects.

This is a simple illustration of a Sequence diagram, as the procedure of resetting the station is rather trivial.

The next illustrations, however, show the principles of the procedure of a normal process. These diagrams are rather huge and may seem unclear on the first view, above all in comparison to the illustrations for the stopping or resetting procedure. If Activity diagrams getting confusing, multiple diagrams should be created for describing the process. Therefore, one creates a less detailed diagram and refines the particular activities with further diagrams. In the example, however, the diagram is just sufficient for its purpose.

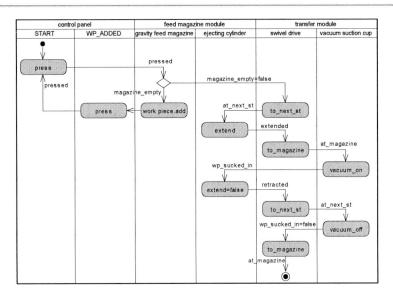

Figure 25: Activity diagram of the Use case normal process.

Figure 25 shows the Activity diagram of a desired procedure of the normal process of the first station. An additionally applied diagram element is a decision node allegorized as a square. This diagram as well shows very intuitively the desired interaction of the actuator and sensor values. Again, the diagram is divided into activity partitions for the single modules and their sub-modules. The operator first presses the START button. If the gravity feed magazine is empty, work pieces need to be refilled, and the button WP_ADDED needs to be pressed, followed again by the START button. Now the swivel drive moves to the subsequent station, and the ejecting cylinder must extend to push a work piece into the delivery position. The swivel drive must now move back to the magazine and takes over a work piece by enabling the vacuum suction cup. Once the work piece is sucked in properly, the cylinder moves back into the initial position, and the swivel drive moves to the subsequent station, where it releases the work piece and moves back to the magazine.

Although not applied in the examples, a fundamental advantage of Activity diagrams is the possibility of describing parallel activities described in [28] and also shown in [24]. A fundamental disadvantage, however, is that it quickly gets semantically overloaded if too many different diagram elements are used as also shown in [24].

Figure 26 illustrates in form of a Sequence diagram the desired procedure for a normal process. This diagram again shows the procedure for the control levels and has pseudo algorithms on the left side derived from the Activity diagram for that particular Use case.

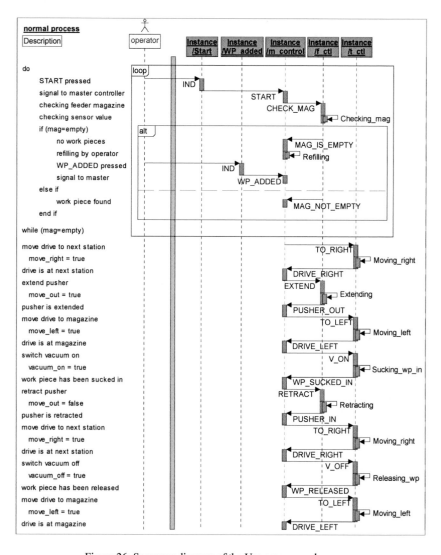

Figure 26: Sequence diagram of the Use case normal process.

Additional elements used in this diagram are loops (*loop*) and decisions (*alt*) to illustrate iterative procedures and sequences, which depend on certain Conditions. These elements are shown as frames around the desired sequences.

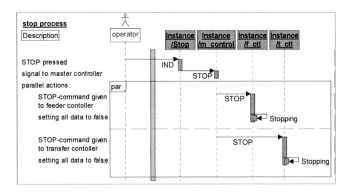

Figure 27: Sequence diagram of the Use case stop process.

The Activity diagram shown in Figure 27 illustrates the procedure of stopping the plant. This diagram is shown in this work, as the possibility of parallelism is used here. If the operator presses the STOP button, the plant is supposed to stop entirely. Therefore, the coordinator sends STOP Events both to the feeder and the transfer unit. This is illustrated with a frame as well. The sequences within this frame, but separated by a dashed line are supposed to execute in parallel. To show parallelism in Activity diagrams, the usage of splitting and synchronization elements is needed as described in [28] and [24].

The specification of the functional behavior of the plant and the controllers is solved with State Machines. The State Machines used within the SysML are Harel State Machines, originated in [29] by David Harel. In this work, however, the State Machines are constructed in a way that they correspond to the Execution Control Charts (abbr. ECC) described in Chapter 3.2, as it still is desired to generate Function Blocks within this Framework. That means, not all of the elements that are provided by this kind of State Machine are used, but only the elements that correspond to an ECC.

This part of the chapter is separated from the other behavior descriptions, as for a specification it usually is not common to create State Machines for the plant elements and their controllers. It is needed for the simulation part to test if the specification illustrated so far is consistent. Actually, defining State Machines of control objects is rather a part for developing engineers who actually program the controllers of plants according to a given specification.

The specification created with the elements so far should cover the mentioned problems that occur with common specification means.

Based on the structural and behavioral descriptions that have been created so far, and eased by the fact that they are available, now a qualitative specification of the plant as well as of the controllers can be created. The blocks created in Chapter 3.4 are now extended by associated State Machines, which describe the behavior of the blocks. For the first module of the distribution station, one therefore extends the blocks *ejecting cylinder* and *gravity feed magazine*. The associated State Machines for the ejecting cylinder and the gravity feed magazine are illustrated in Figure 28.

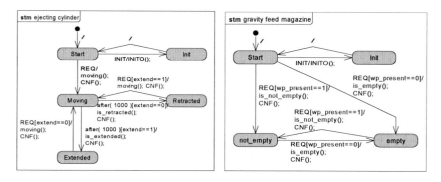

Figure 28: State Machines of the ejecting cylinder and gravity feed magazine plant elements.

These State Machines describe the behavior of each plant element in a way, which it is expected to behave according to certain control signals. After an initialization of the block and its State Machine, the ejecting cylinder can have three states, namely *Retracted*, *Moving* and *Extended*. The transitions between the single states are designed in a way of *Trigger [Condition] / Actions*. In the example of the transition between the states *Retracted* and *Moving*, this means exactly that the trigger is the incoming Event *REQ* associated with the Condition *extend==1*, and then the actions *is_retracted()* and *CNF()* occur, whereas *is_retracted()* is an algorithm and *CNF()* an outgoing Event. Inside the algorithms that are executed, only Data values are set according to the reached state. There can also be other kinds of triggers as incoming Events. For instance the transition between the states *Moving* and *Extended* is designed in a way that the trigger is a time period of 1000 ms, shown as *after(1000)*. The State Machine on the right side of Figure 28 is designed similarly. After the initialization of the block and its State Machine, the gravity feed magazine can have two states, namely *not_empty* and *empty*.

The State Machines in Figure 29 are to be interpreted accordingly.

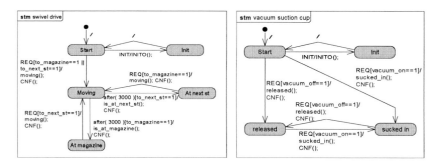

Figure 29: State Machines of the swivel drive and vacuum suction cup plant elements.

The next step is to design the controllers of the first station in a hierarchical way as described in Figure 21. It is started with the task control blocks *f_control* and *t_control* defined in Chapter 3.4 and illustrated in Figure 21. The State Machine of the feeder controller is illustrated in Figure 30.

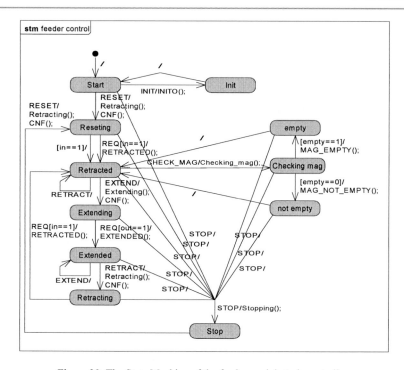

Figure 30: The State Machine of the feeder module task controller.

This controller controls the functionality of the first module of the distribution station, namely consisting of the gravity feed magazine and the ejecting cylinder. It is important to mention that this controller is only connected to the plant module itself via Event- and Data-connections and intercommunicates with the coordinator control element only by sending and receiving Events. The according State Machine of the controller of the transfer module is illustrated in Figure 31.

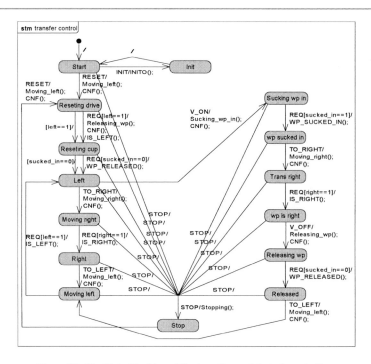

Figure 31: The State Machine of the transfer module task controller.

This controller controls the functionality of the transfer module and is connected only to the plant module as well via Event- and Data-connections and also intercommunicating with the coordinator control element by sending and receiving Events. Both of the task controllers handle algorithms, which assign desired Data values only to their associated plant elements. The coordinator control element instead only sends and receives Events to/ from the task controllers. It does not handle functional algorithms, but is just a controller to coordinate the different activities of the modules. The State Machine of the coordinator control element is illustrated in Figure 32.

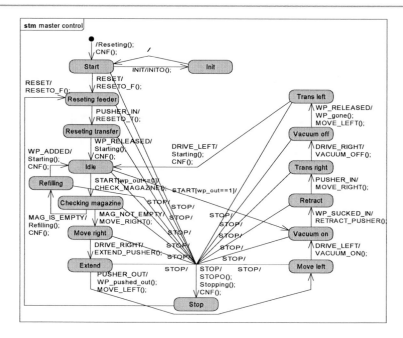

Figure 32: The State Machine of the distribution station coordinator control element.

After all, the description of the behavior of the example automation system is completed. There are functional (State Machines) and non-functional (Activity diagrams, Sequence diagrams) descriptions of the desired behavior, the single plant elements and their functionality. The next step of simulation can now be performed as described in the next chapter, and also, Function Blocks now are automatically generated.

From this stage, then, there already would be executable control code according to IEC 61499 available to control the plant as specified. Since the code is not verified yet, the application of this procedural method would bring a stable control program as simulated in Chapter 3.6, but in all likelihood complications that are difficult to predict in advance are going to appear as usual.

3.6   Simulation

By using appropriate tools, e.g. [3], [4], there is the possibility of simulating the specification. Therefore, with the blocks associated State Machines are executed and tested against each other according to the previously defined Activity diagrams and Sequence diagrams. Therefore, the development environment generates executable simulation code, and the developer has the possibility to simulate several scenarios according to the specification. Above all, in bulky projects this holds several advantages, as already in the phase of creating the specification it can be tested on consistency. The scenarios can be simulated either step-by-step or automatically, without intervention of the developer. In this work, the simulation of the specification was successful for the Use cases, and the specification therefore is usable for model-based design of the controllers [24]. As a simulation of the system again in the IEC 61499 chapter is performed and simulation is only a can-be-performed-part. Now, the IEC 61499 controller generation is described and then the intermediate results of this chapter are pointed out.

3.7   IEC 61499 Controller Generation

For the automatic generation of IEC 61499 Function Blocks, certain modeling guidelines need to be applied, which can be comprehended in the last subchapters. In this chapter, a survey of the mapping between SysML and IEC 61499 elements is given along with a short introduction to the transformation tool, as the transformation is just a secondary issue in this work. It is intended to generate complete IEC 61499 system configurations with applications of Function Blocks. Table 3 shows the mapping between SysML elements and IEC 61499 elements, that is introduced in this work.

Table 3: Mapping between SysML and IEC 61499.

SysML	IEC 61499
Block diagram	System
Block diagram	Device
Internal block diagram	Resource
Internal block diagram	Network of Function Block instances

Block enhanced with a State Machine	Basic Function Block
Operations	Event I/Os and algorithms
Attributes	Data I/Os and internal variables
State Machine	ECC
Block enhanced with Internal block diagram	Composite Function Block

The portability of IEC 61499 Function Blocks is based on unified XML (abbr. Extensible Markup Language) descriptions. The transformation tool therefore features the following functionalities. It imports the XMI-file (abbr. XML Metadata Interchange) that has been generated by the SysML development environment. In this work, the ARTiSAN development environment [3] has been used, which generates one file for the entire system specification. The XMI-file is parsed and the information that is relevant is filtered and converted. Then, new XML files are generated for each specified Function Block, Adapter Interface, system, etc. These files are further augmented with generic IEC 61499-specific information. The tool therefore features the following class hierarchy, which is shown in Figure 33.

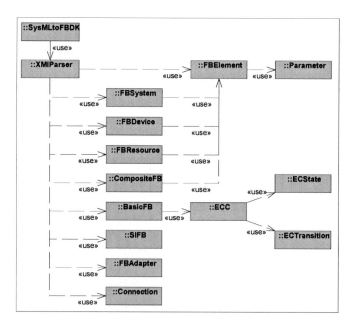

Figure 33: Class hierarchy of the transformation tool.

The Graphical User Interface (abbr. GUI) of the tool is shown in Figure 34.

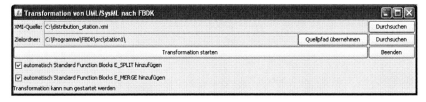

Figure 34: GUI of the transformation tool.

The generated XML files are imported by an IEC 61499-compliant development environment. In this work the import has been tested with the Function Block Development Kit [33] and the O³Neida workbench [32]. The imported Function Blocks then need to be compiled and finally, the executable control specification is available.

The generated control specification in the example of the distribution station (testbed description see Chapter 2.5.1, specification see previous subchapters) is shown in Figure 35.

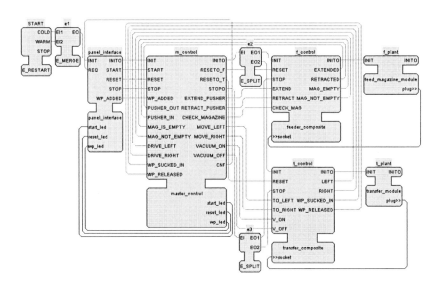

Figure 35: The generated system configuration.

The transformation tool has been developed during the making of the diploma thesis of A. Helbich, which has been supervised by the author.

3.8   Evaluation of the Intermediate Results

In the previous chapters, an entire process could be followed of specifying an automation system including textual descriptions of the requirements, graphical descriptions of the structure and behavior of the automation system. It was anticipated and intended to draft the specification with as much as possible diagrams and other description means but as less as possible textual descriptions, to show the difference between this kind of specification and traditional specifications.

But the application of SysML with highly expensive tools does not only consist of just drawing diagrams. By defining appropriate profiles and by consistent modeling, the functionality of the diagrams can be very useful in huge control software projects. The possibility of the simulation of the specification can gain essential advantages, as already in the phase of specification one can avoid abrasive errors. An outstanding effect of simulation using this description means is, besides the checking of consistency of the specification, the possibility to touch the requirements. With this method, it can be checked if each requirement is concerned.

The usage of SysML is not the solution for all problems that occur at model-based controller design. However, it can ease some of those problem solutions and help to embed model-based design methodologies in a broader context of embedded systems. The structure diagrams are useful for a quick survey of the structure of the plant elements and can help to identify their interactions. These interactions are qualitatively described in the behavior diagrams.

The behavior diagrams differ from each other remarkably, as they describe the interaction between hardware objects as well as interactions of software objects. The functionality of single objects on the other hand is qualitatively described with State Machines.

The application of such a form of specification requires a careful choice of diagrams and diagram elements as well as a conscious abandonment of certain description opportunities. The SysML probably does not have the potential to replace classical engineering methods, but it can definitely ease them. In this work, certain adopted and self-developed modeling guidelines were followed to use this specification to generate a yet executable system specification according to the IEC 61499 standard as described in the next chapters.

4   Executable System Specification

4.1   Model-View-Control Design Pattern

To earn most of the advantages of the IEC 61499 standard, it was decided to use the Model-View-Control design pattern, which has shown up to be a powerful approach in some previous works [15], [16] and is schematically constructed in Figure 36 under application of IEC 61499 Function Blocks. The main idea behind this design pattern is to create first a model of the uncontrolled, physical behavior of the plant, then to build a visualization for this model and then to build a controller, which controls the model (and later on the real plant).

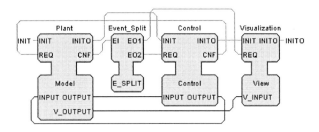

Figure 36: Schematic construction of the Model-View-Control design pattern.

The Function Block named *Plant* represents a model of the uncontrolled plant. This model is informal and contains descriptions of the structure as well as of the behavior of the plant. Therefore, these Function Blocks contain Execution Control Charts (abbr. ECCs), which are Event-driven State Machines and describe just that behavior.

The behavior of the plant can be modeled either discrete-event (for later verification) or continuously (for the visualization output). The ECC supports both of these modeling forms, due to the kind of application of the integrated algorithms.

The FB named *Control* represents the controller of the plant model. It is connected with the plant model in closed loop. This FB schematically represents either centralized or distributed controllers of the plant. It also contains an ECC with internal algorithms.

The *Visualization* FB represents the graphical output component of the system configuration. With adequate means, one can design a sound visualization [15].

For the implementation of IEC 61499 Function Blocks, there are some tools available [30], [31] and [32]. A good solution is the usage of the Function Block Development Kit [33], which has proven to be well applicable in this context [15].

Adapter Interfaces are a useful means for the description of the interfaces of mechatronic components. They explicitly describe the structural appearance of the plant. It is useful to subdivide the plant into several sub-modules, on the one hand to keep the single plant elements separated from each other and on the other hand to ease the process of designing distributed controllers, as the plant appears distributed in that way.

Adapter Interfaces are a reasonable possibility to help in that matter. Once the plant is separated into its sub-modules, one begins to design the Adapter Interfaces according to the interfaces of the sub-modules. One thereby gains a structural description of the plant elements.

Adapter Interfaces further can help to keep system configurations clearly arranged. Therefore, Adapter Interfaces can appear either as plugs or sockets. The idea is the following. Each Model-View-Control element is connected to the according Adapter Interface of the plant sub-module. As the input of a controller is the output of the plant, Adapter Interfaces sometimes need to be mirrored according to their in- and outputs. In case of connecting an Adapter Interface to a controller, it will appear in the same way as the plant and therefore will be called a socket connection. In case of connecting an Adapter Interface to a plant model element, the adapter is mirrored and now appears as a plug connection. The software component of the adapter but still is the same.

In Figure 37 one can see a schematical demonstration of the usage of Adapter Interfaces. A deep description of the usage and background of Adapter Interfaces can also be found in [13].

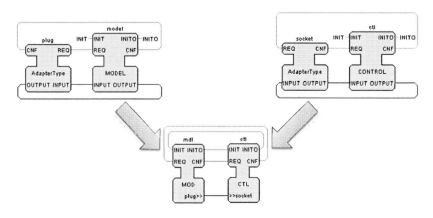

Figure 37: Usage principles of Adapter Interfaces.

4.2   Modeling

The model of the plant represents the uncontrolled, physical behavior of the plant. One should differentiate between two kinds of models. The first kind is a continuous, discrete-event model and the second kind is an only discrete-event model. For a sound visualization, one should prefer the first kind of model. This model not only describes the end positions of mechatronic components and the state that it is on its path during the end positions, but also delivers short interval Event information according to the path of the dynamic component. It therefore publishes short-time refreshed Events to the visualization component. The visualization component can interpret those Events (with according Data) as path information and delivers therefore a smooth motion of the visualized mechatronic component. This issue will be adressed in Chapter 4.3.

At first the kind of discrete-event modeling is described.

This kind is only discrete-event and should serve for verification and predictive control tasks. It therefore usually only publishes Events according to the end positions of a mechatronic component. This kind of model is similar to the models described in Chapters 3.4 and 3.5.

As an example, a typical element of the material handling and processing sector is taken, a *pneumatic cylinder* with a steep return clip as illustrated in Figure 38.

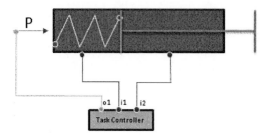

Figure 38: A typical pneumatic cylinder element.

This component is equipped with three Boolean in- and outputs and an embedded task controller. The functionality of this pneumatic cylinder component is to extend the piston until it reaches its end position, and then to retract it back to the starting position.

Figure 39 shows the structural model of that component in form of an Adapter Interface (see Chapter 4.1). It represents the inputs and outputs of that component.

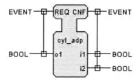

Figure 39: Adapter Interface of the cylinder component.

In Figure 40 one can see an implementation of a discrete-event model of that simple mechatronic component including its ECC. The algorithms *retracted* and *extended* only contain value assignments according to the states (i1=1; i2=0 respectively i1=0; i2=1). As can be seen, besides the state *START*, the ECC contains four further states, namely *retracted*, *move_fore*, *extended* and *move_back*. As mentioned above, an Event-output (*CNF*) only occurs when the states according to the end positions are reached, and the executed algorithms have set the new Data output values.

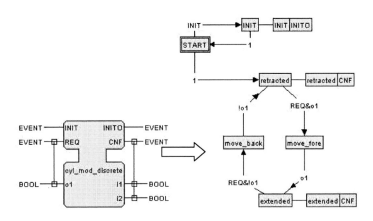

Figure 40: The Basic FB and its ECC of the discrete-event model of the cylinder.

After initialization, the model of the mechatronic component is in the state *retracted*. Once the Event *REQ* occurs along with a Data value o1=1, the ECC will move into the state *move_fore* and then directly under the Condition o1=1 to the state *extended*, where a *CNF* Event is emitted. Once the REQ Event occurs again along with a Data value of o1=0, the model will move to state *move_back* and then directly to the state *retracted* along with an emission of a *CNF* Event, where the path can start over in case of an occurring *REQ* Event. That is a very simple discrete-event model implementation.

Figure 41 shows the connection of that discrete-event model with the Adapter Interface of the mechatronic component shown in Figure 39. In that situation, the adapter of the component is called a plug. In a system configuration, the model will appear as shown on the left side of Figure 41 and can in turn be connected to the controller element with the adapter connection.

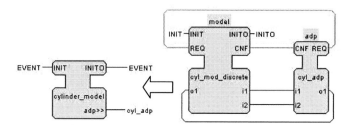

Figure 41: The Composite FB of the model and its internal.

Now, a discrete-event model of the mechatronic component is available. Connected to a controller, it is possible to simulate the behavior of the model and the controller in closed loop. For verification issues, this kind of model is also valid, as described in Chapter 5.

In [16] and [35], the idea to execute models in parallel to the controller of the plant to obtain information for predictive control algorithms is described. In case the real plant begins to deviate from the behavior of the models, those algorithms could analyze it and carry the plant into a safe state or at least, report the deviance to the control engineer. Further investigation in that field also is anticipated by the author. For that issue, continuous kinds of models need to be constructed which do not only describe the discrete-event character of mechatronic components, but also have a quantitative character which shows up in either time-interval or counting methods, that correspond to the real plant.

A kind of continuous modeling is described in the following. The used plant element example is the same as it was used for discrete-event modeling.

While modeling the mechatronic component in a discrete-event way, for a continuous model the path between the end positions needs to be subdivided into several intervals. As one possibility, the entire path is defined as a 100 % distance, and this distance is subdivided as desired and described in the following.

First of all, the discrete-event model interface is taken and augmented with integer values speed input and pos output and further with the Event input *MOVE* and the Event output *STEP* as shown in Figure 42. The new attached Events are associated to the corresponding new Data values.

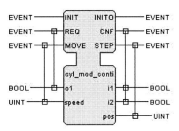

Figure 42: The interface of the continuous model of the pusher.

The Event output *STEP* will induct the visualization component, and the pos value will deliver the very same visualization component with the needed information concerning the position of the piston of the mechatronic component. The distance that the piston will cover is defined by assigning a certain percentage value to the Data input speed and has to be in due proportion to the real mechatronic component. This can be either a 1 % step or another step sized as desired.

The ECC of the continuous model now is described in the following.

The ECC of the discrete-event model in Figure 40 is taken and augmented with algorithms and state transitions as shown in Figure 43.

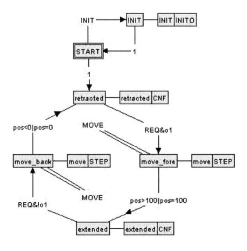

Figure 43: The ECC of the continuous model.

The model of the mechatronic component now again has four states, namely *retracted*, *extended*, *move_fore* and *move_back*. The ECC is extended with the algorithm move. The mode of operation is the following. After initializing the ECC, the model is in the state *retracted*. The pos value is initially set to pos=0. When a *REQ* Event occurs and the Data value o1=1 is true, the model is now in the state *move_fore*. The algorithm *move* is now executed. Each time a *MOVE* Event occurs, this algorithm increases the initial pos Data value (pos=0) about a value of 1. Then a *STEP* Event is emitted. As long as the Data input value o1=1 is set and a MOVE Event occurs, this algorithm is executed. If the pos Data value has reached pos=100 respectively pos>100 the *extended* state is reached. The algorithm works contrary in case the Data value o1=0 is true along with an incoming *REQ* Event. Now, a model that incrementally increases the pos value respectively decreases the pos value is available. To adjust the behavior of the model, a timer Function Block needs to emit an Event to the model (Event *MOVE*) according to the real behavior of the mechatronic component as illustrated in Figure 44, with a fictional value of t#100ms.

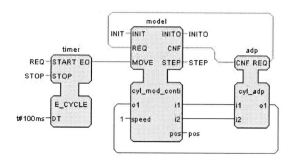

Figure 44: The continuous model with its adapter and a timer FB.

The author is aware of that this solution works fine with visualization components during model-based controller design but probably not for predictive control algorithms, as the interval that the timer Function Block emits the Event depends on the speed of the operating system of the controller respectively. In case it is a real-time operating system, this solution could work approximately, but for hard real-time constraints this is rather not a possibility.

4.3   Visualization

The Model-View-Control (abbr. MVC) approach as the next step includes the building of a visualization, i.e. a graphical representation of the plant and its components. The approach itself is illustrated by several examples provided with the FBDK, and also in [13], [15], [16] and [56] some results for a sound visualization have been presented. In the MVC pattern, the plant is first modeled and visualized, then simulated, then the controller is tested, and later the model is substituted by interfaces to the real plant. This work follows the MVC pattern, so first the models of the plant were built, which describe the uncontrolled, physical behavior of the objects. It was dealt with discrete modeling as well as continuous modeling. Both kinds are appropriate for visualization. Now these models are executed at the engineering and simulation station (which is shown in Figure 2) along with extra defined visualization Function Blocks, which interpret the information concerning the position of a module and, e.g., its end positions. In the example of the distribution and testing station, the result of this procedure delivers a visualization output as shown in Figure 45.

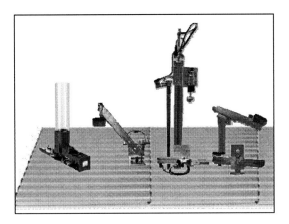

Figure 45: Visualization output.

The visualization works as follows. Looking back to Figure 36, a non closed-loop (unidirectional) adapter connection between the model of the plant (*V_OUTPUT*) and a visualization component (*V_INPUT*) can be seen. Such a *passive* adapter connection only delivers information concerning the current position of a model or its end positions to the visualization component in form of integer Data values. The visualization component usually works as follows.

Pictures (.jpg, .gif, etc.) are assigned to the incoming Data of the model Function Blocks. These pictures are assigned to an initial position (coordinates) in the visualization output frame. Furthermore, it is configured which degrees of freedom the particular components have. Then the models are executed, and the visualization is tested along.

Before the next step, the building of the controller(s), is performed, there is also a need for a Human Machine Interface (abbr. HMI). The real plant normally provides a real interface with buttons, lights, sounds and so on. During Software In the Loop simulation (see Chapter 4.5), however, this interface needs to be built before. In case of using touch screens as a real HMI later or in case there is even no HMI provided by the plant itself (as suggested in [16]), it is indispensable to build an HMI. In [16] as well, there is an example implementation of an HMI-frame as shown in Figure 46 for a car seat controller. Each button there is realized in form of a Function Block, which emits an Event in case of that it is pressed which is then delivered to the control device.

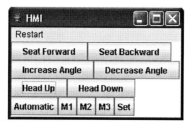

Figure 46: An example for an HMI.

4.4   Systematic Model-Based Controller Design

4.4.1   Distributed Control – A Challenge in Design

There can be many advantages of using distributed controllers. It reduces the complexity of single controllers by splitting the functionality into smaller, functional units, thus saving hard- and software resources. This can bring advantages in the design and maintenance of the controllers. Above all, for very complex control tasks, the splitting of functionality can bring a better overview. Splitting control functionalities into smaller units also has the advantage, that it eases reconfiguration and, therefore, reusability of controllers. Another aspect is the possibility of having concurrently executed control code, and therefore it gains real parallelism of controllers. The aspect of spatial distribution is possibly the most important plus for, e.g., systems like modern cars. The control tasks can be located near to the actually to be controlled device, which avoids long-term latency periods in real-time control tasks between the controller and the controlled object. Along with adequate communication means, e.g. FlexRay [48], new possibilities rise in the direction of safety-critical driving assistance systems. By distributing control, one also can make the entire system safer, primarily in case of a malfunction or a loss of a controller, the remaining control devices could either overtake its control tasks or, at least, bring the system into a safe state which causes no harm to the remaining plant. Another aspect is the possibility of heterogeneity on the part of the controllers. The hardware of different vendors can be installed, they just need to be able to communicate with each other in a standardized way.

A remaining disadvantage is the increased effort in communication and the effort in changing existing centralized control environments into distributed ones, also to mention the rising complexity to coordinate the distributed controllers.

Apparently, the advantages outbalance the disadvantages.

The IEC 61499 standard has been developed for distributed systems. It shall support the mentioned features above, which are parallelism, decentralism, spatial distribution, interoperability and so on. But how are these features used in the best way? In this work, there are some solutions introduced, which shall guide the developer in designing distributed controllers according to IEC 61499.

There are many questions that need to be answered, before one begins to design a distributed system all over. The most important thing probably is the tracing of information flow between the single components. Some other questions rise before one starts.

How are the components arranged? Does the system have sequential information flow between the components or rather disordered information flow? Do the components need to be turned on/ off if desired, without disturbing the rest of the process? Do the controllers support easy reconfiguration of the system configuration?

These and other questions, like gaining productivity by using distributed controllers, are the actuation for the next subchapters. Some approaches to design distributed controllers are introduced and explained along with an evaluation, where the particular approach best fits in.

In the following subchapters, it is shown how to apply the IEC 61499 standard for independent controller design, evolutionary from centralized to distributed control.

4.4.2   Centralized Control Approach

The *handling station* in the exemplary testbed is controlled by a single Function Block which acts as a centralized, sequential controller (Figure 47). This Basic Function Block represents the inputs and outputs of the entire handling station and includes a sequential ECC with several algorithms. The Function Block *handling_centralized_controller* in this configuration further is connected to Service Interface Function Blocks which make the services of the plant outputs (on the left) and the plant inputs (on the right) available.

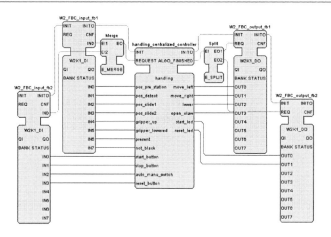

Figure 47: The central controller connected to Service Interface Function Blocks.

The purpose of controlling the station with a single FB is to show the migration of IEC 61131 compliant algorithms such as Structured Text [8] into an Event-driven architecture based on IEC 61499. Therefore, the control Function Block provides the Event input *REQUEST* and the Event output *ALGO_FINISHED*. Each time the *REQUEST* Event occurs, an image-refresh of the associated inputs is supplied to the internal ECC, which executes the dedicated algorithms. If the execution of an algorithm inside the ECC terminates, all associated Data outputs are refreshed, and the Event output *ALGO_FINISHED* is triggered. The ECC represents a transformation of a well-known IEC 61131 compliant Sequential Function Chart, but without parallelism. To this end, the controller just represents a transformation of a synchronous PLC-based control into an asynchronous Event-triggered control architecture.

If one implements control using this strategy, the advantages of the IEC 61499 standard are used by no means. It is not necessary to examine the information or material flow of the plant, which is therefore not examined graphically in this subchapter. The controller represents a non-reusable, inflexible, sequential way of implementation. The controller could only be reused in the very configuration of this plant. The next subchapters deal with distribution of controllers, always in context of the exemplary specific plant, but always fully reusable and therefore flexible.

On the other hand, this way of implementation allows for using already existing, IEC 61131 compliant controllers, for instance when integrating IEC 61131 into IEC 61499 control environments or the other way round.

4.4.3   Purely Distributed Control Approach

The first approach of distributed control that this work deals with consists of purely distributed autonomous controllers with no coordinator. This approach considers a sequential information flow or material flow between the single components, so that the engineer faces a line of controllers. Material as well as information flow are illustrated in Figure 48 as follows.

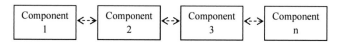

Figure 48: Sequential arrangement of mechatronic components.

This arrangement allows designing control objects in the example of the *distribution station* as shown in Figure 49. The main idea is that each object performs its control actions, along it does not get in conflict with another object, namely by clashing of the objects.

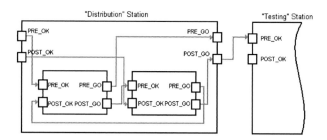

Figure 49: Signal Interface of control objects in the external blocking approach.

This idea leads to a Function Block implementation, which is reusable and fully distributed as illustrated in Figure 50.

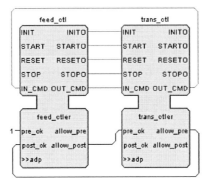

Figure 50: Implementation of the purely distributed autonomous control architecture.

That situation permits the design of standardized Event and Data inputs and outputs of the controllers. Therefore, some architecture levels of distributed controller design need to be described. On the highest level, the controllers provide Event inputs *START*, *STOP* and *RESET*. To connect Events through in the sequential order of controllers, they furthermore provide Event outputs *STARTO*, *STOPO* and *RESETO*. Furthermore, the controllers provide information output *allow_pre* and *allow_post* associated with an Event *OUT_CMD* and information input *post_ok* and *pre_ok* associated with an Event *IN_CMD*. On that level of controller design, the engineer does not have to consider pure control functionalities but the order of the controllers. One level beneath, the controller FBs are connected to the interfaces (in the example in form of Adapter Interface Function Blocks) of the plant and contain control functionalities of the single modules. They also provide information of the state of the controller. In particular, the controllers contain the control functionalities of just one module and do not care about the control functionalities of their environment. The big advantage of this design approach is the reusability of the controllers. They could be removed from that "chain arrangement" and for instance be inserted into another position. Other components could be inserted into the system configuration as desired. This architecture was introduced in [36] and [15]. Up to this arrangement of controllers, it is obvious to design the controllers in a way as proposed in this chapter. In the next chapter though, the plant is configured in a different, more complex way, which forces the control engineer to apply other design approaches for distributed control.

4.4.4   Coordinator Control Approach

This work focuses on industrial automation and mainly on components and their reuse [43]. Industrial automation systems usually consist of numerous components, each with its certain functionality. To control each of these components, it is proposed to develop so called *task controllers*, which control only the specific functionality of the component [36].

It is important, not to mix the *task concept*, as popularly known from real-time operating systems in the embedded system sector (e.g. OSEK [37]), with the task controller concept explained in this work.

As an example, a typical element of the material handling and processing sector, a *pneumatic cylinder* with a steep return clip is taken. The models of these components are obtained from Chapter 4.2 and again illustrated in Figure 51.

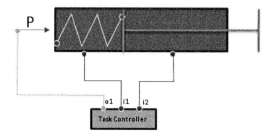

Figure 51: The pneumatic cylinder element with task controller.

This component is equipped with three Boolean in- and outputs and an embedded controller. For the embedded task controller of this component this implies the output *o1*, which represents the input of the plant component, and the two Boolean inputs *i1* and *i2*, which represent the outputs of the plant component.

The functionality of this pneumatic cylinder component is to extend the piston until it reaches its end position, and then to retract it back to the starting position.

In Figure 52 one can see the task controller with coordinator pattern. The figure is explained as follows.

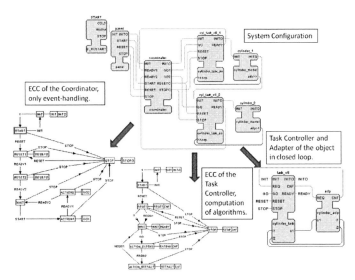

Figure 52: Task controller with coordinator pattern principles.

The System Configuration shows a set of two cylinder elements, illustrated with two identical models of the cylinder (FBs *cylinder1, 2*) and two equal task controllers (FBs *cylinder_task_ctl1, 2*), all represented by Function Blocks. Furthermore, one can see a coordinator Function Block (FB *coordinator*) for this configuration and one Function Block representing a Human Machine Interface (FB *panel*).

Down right in Figure 52 one can see the task controller connected to an Adapter Interface of the cylinder module, represented by an Adapter Function Block. This kind of Function Block usually represents the static structure of components in a system and serves as an interface connection between the Model-View-Control Elements.

The task controller is designed in a way with a unified standard interface, representing the functionality of the cylinder with an Event input *GO* and other standard Event inputs *RESET*, *STOP* and *INIT*. The Event outputs for this task controller are *INITO* and *READY*.

The particular feature of the task controller is, that it only controls the specific functionality of the cylinder, not caring about its environment or scheduling of the tasks. It therefore is equipped with an according Execution Control Chart (abbr. ECC). This ECC (down central in Figure 52) defines the actions of the component along with the computed algorithms.

The scheduling and avoidance of clashes of the objects during the operation of the plant is performed by a coordinator. In the example, the operation sequence is as follows. When a

START Event arrives at the coordinator, the first cylinder shall extend and retract, followed by the second cylinder, which has to perform the same operation. Then, the operation terminates, until another *START* Event arrives. It is possible to stop the operation sequence at any time by emitting a *STOP* Event to the coordinator.

The realization of this operation sequence can be seen in the ECC of the coordinator down left in Figure 52. It is important to recognize, that the ECC of the coordinator Function Block does not contain any algorithms, it only contains Event-handling. The according algorithms are computed by the task controllers.

The author is aware of the triviality of this example, but it serves for easy comprehension of the concept.

The advantages of this control design pattern are obvious. By splitting control functionality one reduces the complexity of the controllers. This implies easier verification as described and shown in Chapter 5. Furthermore, this pattern naturally enhances the reusability of the controllers and eases the reconfiguration of systems.

The coordinator is developed in a way that it can be generated from the specification developed in Chapter 1. Also its verification is eased by the fact that only operation sequence definitions need to be verified. The verified task controllers can then be considered as black boxes during the verification of the coordinator, as they have been verified yet.

As in the approach in Chapter 4.4.3, every module of the *testing station* has its own controller as shown in Figure 54. It is not possible to adopt the approach from Chapter 4.4.3, as there is no sequential information flow for that station of the testbed. The information flow in case of the *testing station* is illustrated in Figure 53. This simply derives from the purpose of the station as described in Chapter 2.5.

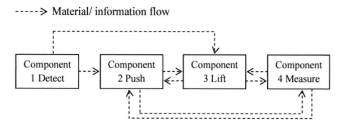

Figure 53: Disordered information flow in the case of the testing station.

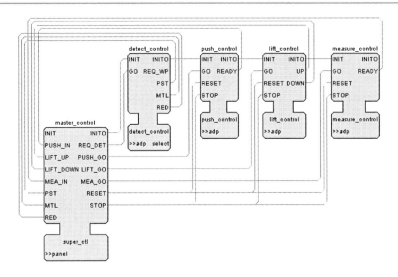

Figure 54: Implementation of the coordinator control pattern.

It is intended to have the possibility to reconfigure and reuse these controllers in any other system by picking them out of a library and place them into a new system configuration. So the controllers also need to have standardized Event inputs and outputs according to their states. The standardized Event inputs are *GO*, *RESET* and *STOP*. The Event outputs of the controllers fit to the state transitions of the controlled objects. The control logic is captured in the Execution Control Charts of the corresponding blocks. The controllers of the modules, that perform certain tasks like "push out" or "measure" trigger a *READY* Event once the desired procedure is done. The master controller has the same Event input names as the distributed controller's Event outputs. The master controller receives Events of the distributed controllers, and according to its ECC, the whole process is coordinated. It is easier to exchange, add or remove single modules in that approach, as distributed controllers are reusable due to the standardized Event in-/ output architecture and are not bound to the organization of the plant. Another advantage of that approach is that a distributed controller can be tested and verified in an easier way as it only includes the ECC of one module. The master controller has no algorithm inside; it coordinates the whole process only with its ECC. By means of that control approach, it is easy to reconfigure the plant as desired. Only the ECC of the master controller must be adjusted or easily rebuilt, and the system engineer has not to think about internal algorithms for the functionality of the single modules.

4.4.5   Distributed Coordinator Control Approach

In the previous chapters, particular design methodologies of distributed control (Chapter 4.4.4) could be seen which enhance flexible reconfiguration and therefore offer reusability of the controllers [36], [15]. Reusability means that control program Function Blocks can be used in different contexts and for different systems. Such programs are restricted to some constraints. An example for a design methodology for components with high reusability was the design of distributed controllers with a standard interface as described in the previous chapter. In Chapter 4.4.2, all controllers are designed with the same predefined interface and are therefore combinable as desired. But it could only be used for *sequential* information and process flow. The applicability to different contexts is thereby restricted.

The solution in this work for the conflict between flexibility and reusability is to split the controller into a reusable part and into an individual system fitted element. Executable sub-processes defined by the physical plant elements called tasks on the one hand and the conditions defining when these tasks have to be activated on the other hand are qualified. So a reusable *task* layer and a flexible *coordination* layer are defined. The combination of reusability and flexibility is a property of the master control design pattern [15] and, equally, of the distributed controllers as described in that work.

A further step is to subdivide the control architecture into *multiple layers* as diagrammed in Figure 55, and to define standardized interfaces for interaction of the control objects between the control layers. The information flow between the objects in the superposed layers is bidirectional. The information exchange between the plant modules symbolizes the modeling of the work piece flow and the modeled properties of the work pieces.

The approach of the master controller design pattern based on [36] is therefore enhanced by splitting this controller into multiple master controllers for each module. So, the result of this approach is a composition of distributed controllers over two layers and a physical plant layer. The physical *plant layer* includes either the models of the plant (as described in Chapter 4.2), or the real plant, work piece flow and work piece properties. The models of the plant are later substituted by appropriate Service Interface Function Blocks [15] to access the resources of the real automation system.

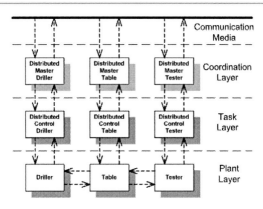

Figure 55: The multi-layered architecture and the information flow between the several layers.

The *task layer* includes distributed controllers, which control each plant object. These controllers only control the specific object, they do not contain any information of the entire process behavior, as it could be seen in the example in the previous subchapter. Algorithms to control the functionality of the plant modules are part of the Task Layer controllers. The task controllers are associated with the plant module. They only control the specific tasks the module is able to accomplish. They are activated by the Event *GO* from the coordination controller and confirm the completion of the task with a *READY* Event. Thus, a simple standard interface for communication between task- and coordination layer is used. The task controllers are implementable independently of the composed system they will be part of. Together with the elements of the plant layer, the task layer controllers are part of the automation objects.

The *coordination layer* in the presented architecture contains the distributed master controllers which coordinate the interaction of the lower layers. The communication between the master controllers can be realized via any communication medium, i.e. a bus system or wireless. The coordination controllers will be designed for each process, or are automatically generated from (semi-) formal specifications from Chapter 1, as proposed in [38]. The ease for control engineers is significant, regarding the following details.

It is further decided that the coordination layer controllers, in contrast to the task controllers, will only contain Execution Control Charts but no algorithms. The coordination controllers accept information about the progress of the processes of the other system parts by their corresponding coordination controllers. The information is exchanged by means of Events.

Such a control problem is naturally implementable by the state-oriented ECC. The limitation on the ECCs simplifies the verification, as will be shown later on. It is, however, no necessary limitation for application of the verification as proposed in Chapter 5.

The introduced design approach offers the following advantages.

A simple way to implement concurrent behavior is provided.

Controllers can be easily mapped to different resources of a particular control configuration.

According to the restriction on ECCs in the coordination layer, the master controllers only have to handle Event flow and are not additionally loaded with computing algorithms.

Splitting the function of coordination depending to the related task supports reconfiguration. Thereby, the modification of the system makes the design engineer able to focus on the modified system part.

And, last but not least, a clear and concise system engineering methodology, in contrast to a purely intuitive one, eases specification and formal verification of the design.

The drilling station was originally controlled by a standard PLC (operation layout see in Figure 56). This setup was changed, and the plant is controlled with an IEC 61499 compliant device to point out the following issue.

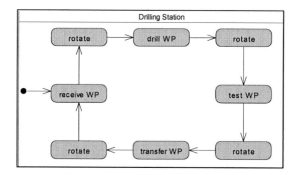

Figure 56: Activity diagram of the sequential operation of the plant.

Because of the physical structure of the plant it is possible to implement concurrent behavior between the modules, i.e. the *drilling module* and the *testing module*. The multilayered architecture that is used in this chapter allows implementing concurrent process behavior in an *easier way* than with a standard PLC device that is programmed, for instance, in Structured Text.

In contrast to the operation layout described in the diagram in Figure 56, the distributed control architecture is used to implement a process that provides concurrency between the modules. This is illustrated semiformally in the Activity diagram in Figure 57. Anytime the table rotated round 90° and in case a work piece is available on the position of the drilling or the testing module, these modules can operate independently. The table module in this case acts as a kind of clock generator. In case there is no work piece available for one of the modules, the *READY* Event triggered by the corresponding task controller confirms this state transition. The reader shall note that the diagram in this chapter is a concurrent behavior description from scratch; it naturally is to be expanded and refined for the aimed automatic generation of the coordination controllers.

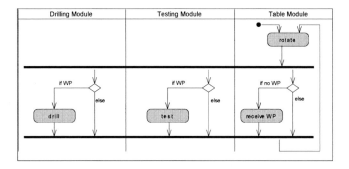

Figure 57: Activity diagram of the concurrent operation of the plant.

Figure 58 shows the implementation of the layered architecture using IEC 61499 Function Blocks in the example of the *drilling station*. The FBs called *drill_plant, table_plant* and *test_plant* mirror the physical, uncontrolled model of the entire plant. That means, FBs are inserted inside those Composite Function Blocks, which represent both the uncontrolled, physical model of one module including the according in- and outputs and the model of the work piece position and properties depending on that plant part. These model FBs are connected to Adapter Interface FBs in form of plugs [13], which also represent the in- and outputs of the modules, but mirrored vertically (see Chapter 4.1). This is done for easier connection between the plant layer FBs and the task layer FBs. The *adp>> / >>adp* edges in

Figure 58 represent *all* Data in- and outputs of the modules. It is necessary to mention that this connection already represents the closed loop of the models and the controllers of the modules, which is *essential* for verification. The FBs *drill_task*, *table_task* and *test_task* represent the task controllers for each module. FBs are inserted inside these Composite FBs which control the functionality of each module. These FBs are also connected to the Adapter Interfaces, but in form of sockets. The *drill_coord*, *table_coord* and *test_coord* FBs coordinate this composition of the task layer FBs and the plant layer FBs. The communication between the coordination layer FBs is realized with the additional Event interfaces *WP_IN*, *WP_O* and *READYO*. The *READYO* Event outputs inform the other coordination controllers about a finished task operation. The *WP_IN* and *WP_O* Event interfaces describe the observed work piece flow between the modules.

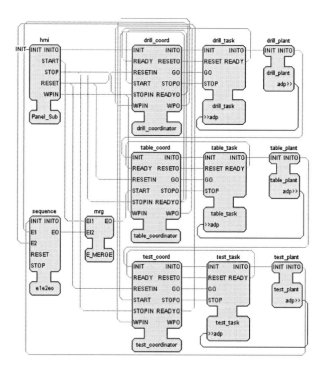

Figure 58: Function Block implementation.

In Figure 59, Figure 60 and Figure 61, one can see the ECCs contained in the exemplary coordinators for the drilling station. The reader shall note that there is no Data computation inside the coordinators by algorithms, but only Event handling to manage the process. It is no surprise that the coordinators in the example look very similar, as the Event interface in this control architecture is standardized and this implicates easy development of the coordinators. It is remarkable that the operation is such trivial. That fact potentially eases the verification approach, which is described in the Chapter 5.

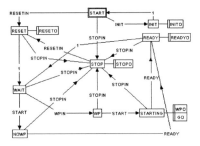

Figure 59: The ECC of the Table module.

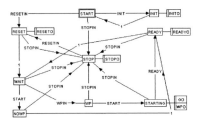

Figure 60: The ECC of the Drill module.

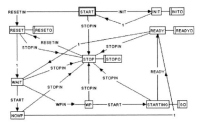

Figure 61: The ECC of the Test module.

4.5   Simulation and Testing

The Model-View-Control framework is completed, and each part is accomplished. Thanks to the models (discrete-event and continuous), the numerous designed controllers and the visualization, it is now possible to perform Software in the Loop (abbr. SIL) tests of the closed-loop behavior of the plant and the controller. Thanks to the yet executable specification according to IEC 61499, the Function Blocks, which describe the system, can be executed. This brings several advantages. Although a consistent specification (see Chapter 1) which helped avoiding errors and misunderstandings during the implementation of the Model-View-Control Function Blocks is available, there is still the possibility of checking the designed system again. For an engineer, it is a big ease if the system that was designed is observable during the simulation and testing scenarios. Wrong implementations of the controllers can be detected by view, which is much easier than working through status reports of the controllers.

In case of the usage of the coordinator control pattern (Chapter 4.4.4), different scenarios can be simulated by just adjusting the coordinator Function Block. This can either be specified as in step one of the engineering framework, and then the coordinator is automatically generated, or it can be done manually in step two. The biggest advantage of SIL testing is that it is not necessary to test drafted controllers with the real plant. That fact saves costs on the one hand, but also partial damaging on the plant or even human beings. In Figure 62, one can see a typical configuration for SIL testing, containing two stations as model devices, a device for the HMI along visualization and, of course, two simulated control devices.

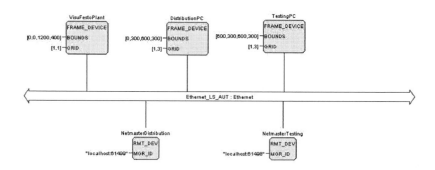

Figure 62: Configuration for SIL testing.

4.6   Evaluation of the Intermediate Results

In this chapter, it was worked over the IEC 61499 standard, dealing with the basic principles, the Model-View-Control design pattern and the simulation of the system with SIL testing. The Event-driven and component-based architecture of the IEC 61499 standard brings several advantages. First of all it supports reusability of components, which could be shown in the chapters of modeling and designing controllers. These components can be reused in other projects with other system configurations. This is one important aspect of the idea of reconfigurability. The models, visualization and control components can be deposited into a library and be inserted into different system configurations via pick and place. The hardware independence of the Function Blocks supports this issue even more. By following the Model-View-Control design pattern it showed up to be a clearly arranged approach ending up in simulation and SIL testing of different production scenarios.

Adapter Interfaces are a useful means to describe the structural interfaces of the single components of the plant. Connecting the Model-View-Control elements with Adapter Interfaces, system configurations are clearer and become not overloaded. Furthermore, Adapter Interfaces offer a useful means for a standardized connection of models respectively Service Interface Function Blocks and their controllers.

The created models can either be discrete-event for later verification issues or continuous for a good visualization. It showed up to be not very difficult to build discrete-event models of the system that describe the uncontrolled, linear physical behavior of the plant using IEC 61499 Function Blocks. These models also can be deposited into a library and later be used for new system configurations. The models also can be executed in parallel to the real plant to perform maintenance or observing tasks.

The largest chapter deals with the design of controllers. Several approaches for distributed control architectures have been introduced and implemented. The most effective control strategy depends on the particular system configuration, above all the material- and information flow between the single components, but in actual fact, the author draws the conclusion, that the coordinator control approach is the most applicable one. The task controllers only need to be developed and later verified once, they can be reused in any other context in each other system configuration. This is a consequence of the Event interface, which was designed consistently. Those task controllers control the functionality of one single component, without caring about production policies or neighbor components. This is then done by the coordinator(s). These Function Blocks do not have to control the sensor/ actuator level in the control hierarchy, they only need to coordinate the sequence of actions of the components, which can be performed in parallel if possible. This is rather easy to read out of specification forms like Activity diagrams and/ or Sequence diagrams from the specification in step one of the engineering framework and rather easy to implement in the coordinator Function Blocks.

Visualization components also need to be designed only once and then can be reused. The visualization is a very useful means when testing the implemented system.

In this chapter, approaches and solutions could be shown that fulfill the postulations of the emerging IEC 61499 standard. The coordinator control along with models and visualization approach fulfills the idea of the component-oriented software for encapsulation of intellectual property. The approach also ensures to have a functional completeness of the system. The possibility of distributing and integrating applications is ensured by the idea of Function Blocks itself. The portability of Function Blocks to different hardware is ensured thanks to a unified XML description, which in turn allows interoperability of hardware from different vendors (shown in Chapter 6). The reuse of components has been realized by the component-based modeling approaches of the models and the controllers, thus leading to flexible reconfigurability of system configurations.

The IEC 61499 standard showed up to be mature for systems engineering. However, the author is aware of certain open points in the standard, which have been considered in numerous publications of distinguished work groups as e.g. in [57], [58], [59] and [60]. These open issues include execution semantics, Event- and Data handling and some other problems, which are not defined precisely in the standard [6] and need to be refined as currently done in the O³Neida work group on Compliance Profile [61].

5   Verification

5.1   Basic Principles

There is the possibility of simulating the designed system during the phase of specification (step one of the engineering framework) and also during the phase of implementing IEC 61499 Function Blocks (step two). But performing simulation is just one step into the direction of a properly working system. A disadvantage of simulation is that only previously defined scenarios can be tested out. States, i.e. combinations of variables that are difficult to predict in advance, often cannot be observed during a simulation of the system. But still, these combinations can occur in the worst scenario. It was walked along a quite robust way of designing distributed systems so far, as well as in control design as in modeling, but a formal verification framework will round out this engineering framework.

There are different understandings for the term of formal verification. On the one hand, there is the engineering science approach. This approach consists of formal models of the plant and the controller in closed loop and delivers a mathematical proof, that the model of the closed loop fulfills a formal specification [68]. On the other hand, there are approaches to formally verify logic circuits, e.g. [67] or formally verify pure code, e.g. by a static analysis [65], [66]. In this work, the first approach is used.

In this approach, on the one hand one can verify production requirements and on the other hand safety/ functional safety requirements. A methodical way how to verify production requirements has been shown in [43], where the production requirements have been formally formulated and verified with formal models of the plant and the controllers in closed loop. The result was a verified system consisting of a composition of Automation Objects [1] which are coordinated in a certain production sequence. In this work, however, it is concentrated on functional safety requirements. Figure 63 shows a cutout of the third step of the engineering framework described in Chapter 0 and illustrated in Figure 1.

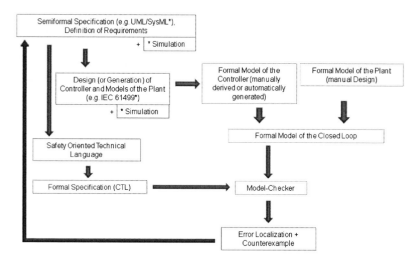

Figure 63: A framework for formal verification.

The illustration shows the required components for a formal verification framework and their relations. Starting from the upper left side of the illustration, it is arranged in the first step of the engineering framework. This step has been examined in Chapter 1. There are two ways from this step. The first way is the automatic generation of IEC 61499 Function Blocks (models of the plant and of the controllers), and the second way is to use the formulated requirements and to transform them into a formal specification. The next way leads to formal models of the controller (manually derived or automatically translated from the control specification) and the formal models of the plant. The formal models of the plant need to be created manually, or are (at least in structure) partly generated automatically [40]. Combined together, these models form the model of the closed-loop behavior of the plant and the controller.

The next step is to feed a model-checker with the model of the closed loop and the formal specification, which contains the functional safety requirements or the production requirements. The model-checker in case of errors computes the location of errors and gives a counterexample. The development engineer then has to rework the specification and the IEC 61499 models, and the process of verification is repeated iteratively.

5.2   Formal Models and Tools

Formal verification always needs a formal model of the plant, which represents the uncontrolled, physical discrete-event behavior. Further, there is a need for a formal model of the controller. The composition of the models of the plant and the controller brings the model of the closed loop. This is indispensable for formal verification, as only the model of the closed loop describes the complete model of the system. For formal modeling in this work it is decided to use Net Condition/ Event systems (abbr. NCES) [14]. NCES are an extension of the well-known Petri nets. A difference is the property of modularity and composition of modules. That means that NCES have a modular hierarchical structure. NCES consist of basic modules and composite modules. The basic modules contain the behavior models while composite ones only contain other basic or composite modules and signal interconnections. Both kinds can have two kinds of signal inputs and outputs. The nets within basic modules contain the Petri net elements places, transitions and arcs extended by Event and Condition signals. Condition signals carry state information and connect places or Condition inputs with transitions. Event signals carry state transition information connecting transitions or Event inputs with transitions. An NCES model example is shown in Figure 64.

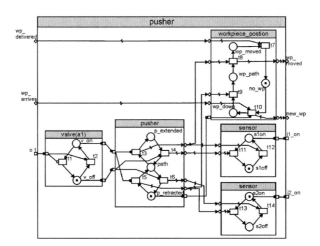

Figure 64: The example of a formal model for the plant element "pusher".

This illustration not only shows a formal model of a plant element, but also indicates the modularity of NCES and the combination of modules. It is the same mechatronic component as described in Chapter 4.2 with the very same interface. The creation/ generation of NCES is performed using a transformation tool developed by I. Ivanova-Vasileva and C. Gerber from the group of Halle [40], [41], [42] or an editor.

A further need for formal verification is a formal specification, which can be used for computing by means of model-checking and is created by following formal rules. In this work, as a formal specification means the Computation Tree Logic language (abbr. CTL) [64] is used. Examples of terms can be found in [43] and in the next subchapter. In this work, it was started gathering the requirements as shown in Chapter 3.3. By formulating the requirements on functional safety in a way like the Safety Oriented Technical Language developed by Heiner, Mertke and Deussen [62], it is possible to automatically generate specifications in CTL by using a tool developed by S. Preusse from the group of Halle [63]. An example of this procedure can be found in the next subchapter.

For model-checking issues, a tool called Signal/ Event Systems Analyzer (abbr. SESA) [44] is used. Figure 65 shows the ingredients that are needed for formal verification using SESA.

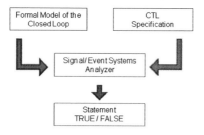

Figure 65: Model-checking with SESA.

A possible error is located, and a counterexample, which does not fulfill the specification, is indicated. SESA therefore shows a trajectory up to the faulty state. This can be visualized in form of Gantt-Charts as well [71]. In case that no error is computed, the model-checker provides the output TRUE.

The modeling technologies and the needed tools are on hands. The next chapter describes an example for the verification framework.

5.3   Verification Example

Under application of the introduced verification framework and by using the mentioned tools and methods, now a formal verification of functional safety at the example of the drilling station described in Chapter 2.5 will be performed. A picture of the real drilling station can be seen in Figure 5.

It is memorized the purpose of the station and its modules. In the context of this chapter, it is focused on the drilling module (2) and the table module (1) as shown in Figure 5. The table rotates round 90° each time, after a work piece is processed and then again. It is obvious for the engineer, that the moving of the drilling module is only allowed in case of drilling of work pieces, which means the drilling module must be turned on. Also, the table must not move while the drilling module is active.

The drilling module has the actuators *extend* and *drill_on*. The sensors of the drilling module are *retracted* and *extended*. The table module has the actuator *rotate* and the sensor *table_positioned*.

Now some exemplary safety requirements for the drilling station are formulated as follows and deposit in a Table as already shown in Chapter 3.3. This is shown in Table 4.

Table 4: Functional safety requirements.

Name	Number	Text
Requirement	1	The drilling module must be activated in case the cylinder moves downwards.
Requirement	2	The drilling cylinder must not move downwards and must be in the upper position in case of the table rotates.

Safety requirement 1 concerns the drilling module and safety requirement 2 concerns the combination of the table and the drilling module.

Those safety requirements are formulated without considering rules or special terms of order of the words.

Now the control architecture of the drilling station as shown in Figure 66 is examined.

The drilling module is connected to a task controller as described in Chapter 4.4.4 as well as the table module.

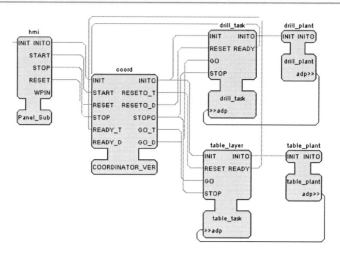

Figure 66: Hierarchic control architecture for the drilling station.

The same way to control the drilling station as described in Chapter 4.4.4, following a hierarchical approach, the coordinator control approach can be recognized. Therewith, a reduction of complexity of the task controllers is ensured, as the task controllers only control the functionality of one module, without caring about the production sequence or even neighbored objects. The production sequence is coordinated by only one Function Block.

The Function Blocks named *drill_plant* and *table_layer* represent the models of the drilling station. The models of the plant are designed in the same way as described in Chapter 4.2. They are connected to the Function Blocks *drill_task* and *table_task*, which represent the task controllers of the modules, in closed loop. The Function Block named *coord* represents the coordinator, connected to the task controllers. Note, that the coordination Function Block is connected with the task controller Function Blocks exclusively via Event connections. This Function Block coordinates the production sequence on a level, where actuator and sensor signals are of no significance. A further Function Block connected with the coordinator is *hmi*, which represents the panel of the plant.

In Figure 67 one can see the interface of the task controller of the drilling module and its ECC. One can recognize that the task controller is able to reset the drilling module and to perform one drilling procedure. Also, there is a state named STOP, which can be reached from every other state. The algorithms associated to the states only contain assignments of variables.

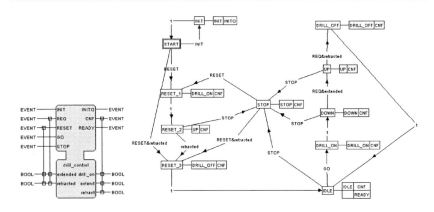

Figure 67: Drill task controller and ECC.

In Figure 68 one exemplarily can see the result of the automatic translation of the drill task controller to a NCES module. One can recognize the same Event inputs and outputs as well as the same Data inputs and outputs. The model inside the module represents the ECC of the task controller of the drilling module.

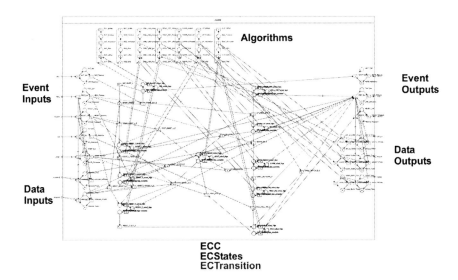

Figure 68: Automatic translation of the IEC 61499 drill task controller to an NCES module.

In Figure 69 one can see the same system configuration as shown in Figure 66 represented with NCES. One has the same components representing the models, task controllers, coordinator and the HMI.

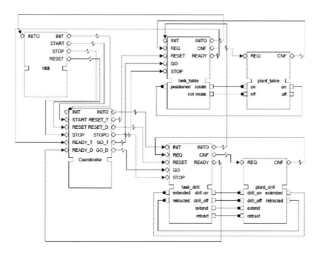

Figure 69: NCES System Configuration.

One now can compute the reachability graph of the drilling station, as formal models of the plant and of the controllers in closed loop are available. This is one of the aspects that come along with formal verification, namely the ability to compute each reachable state of the combination of plant and controller. The result is the reachability graph shown in Figure 70.

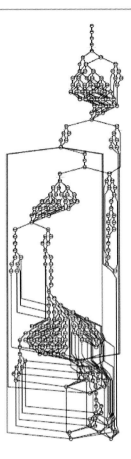

Figure 70: Reachability graph of the drilling station.

This graph consists of 216 states and shows all trajectories that are possible within the models of the drilling station. It is obvious, that a visual verification for that complexity is not feasible. Nevertheless, this computation is used for formal verification.

One now takes the safety requirements previously defined and reformulates them in the Safety Oriented Technical Language.

For the drilling module:

> <<*Requirement 1*>> *"The drilling module must be activated in case the cylinder moves downwards."*
>
> In Safety Oriented Technical Language:
>
> *It holds always : If extend holds , then it holds simultaneously : drill_on .*

And for the drilling module in combination with the table module:

> <<*Requirement 2*>> *"The drilling cylinder must not drill and must not move downwards and must be in the upper position in case of the table rotates."*
>
> In Safety Oriented Technical Language:
>
> *If rotate holds , then it holds simultaneously : not extend and retracted and not drill_on .*

One recognizes that the descriptions of the safety requirements do not vary that much, so it is possible to formulate those requirements in the Safety Oriented Technical Language (abbr. SOTL) right in the beginning of the entire engineering framework, namely in the SysML specification. One now uses the tool described in Chapter 5.2 and generates formulations in CTL based on the SOTL formulations. The results are the following terms.

For the drilling module:

> *(extend)* $\rightarrow$ *((drill_on))*

And for the drilling module in combination with the table module:

> *(rotate)* $\rightarrow$ *((! extend & retracted & ! drill_on))*

One feeds the model-checker SESA both with the model of the closed loop and the CTL safety requirements. The computed result is the following.

For the drilling module:

$$AG \, (\, (\, extend \,) \, \rightarrow (\, (\, drill\_on \,) \,) \,)$$
$$AG \, (\, (\, (\, p14 \, v \, p13 \,) \,) \, \rightarrow (\, (\, p10 \,) \,) \,)$$

Result SESA: TRUE

And for the drilling module in combination with the table module:

$$AG \, (\, (\, rotate \,) \, \rightarrow (\, (\, ! \, extend \, \& \, retracted \, \& \, ! \, drill\_on \,) \,) \,)$$
$$AG \, (\, (\, (\, p18 \,) \,) \, \rightarrow (\, (\, ! \, p13 \, \& \, p15 \, \& \, ! \, p10 \,) \,) \,)$$

Result SESA: TRUE

One can see as a result, that the model-checker did not locate errors and therewith did not give a counterexample. The functional safety requirements therefore are fulfilled.

In case of detected problems, the IEC 61499 Function Blocks respectively the specification need to be adjusted iteratively, and the process of model-checking needs to be restarted.

5.4   Evaluation of the Intermediate Results

As some intermediate results, one can see the following. A framework for formal verification has been introduced, which fits into the engineering framework described in Chapter 0. This framework describes an approach for formal verification, which needs formal models of the plant and of the controller in closed loop. A further need is a formal specification along with an appropriate model-checker. For application of this framework, the model-form of Net Condition/ Event systems has been chosen, which showed up to be very useful in this context. As NCES fit with the Event-based structure of IEC 61499, one can automatically generate formal controller models in NCES based on IEC 61499 Function Blocks. NCES support modularity and composition and therefore match with common concepts of engineers. Due to this property, it is possible, at least in principle, to generate structural plant models based on IEC 61499. The formal models of the plant behavior, however, need to be designed manually or semi-automatically [41], [42].

The framework further describes the result of a formal specification starting from the first step of the entire engineering framework. By applying simple rules of formulation, safety requirements can automatically be generated in Computation Tree Logic based on formulations of the Safety Oriented Technical Language respectively formulations during the phase of specification with the Systems Modeling Language. Therefore, an appropriate tool has been developed in the work group of Halle.

By having on hands formal models of the plant and the controller in closed loop along with the formal specification, the next step is to feed a model-checker. Therefore, in this work the model-checker SESA is used. SESA computes the formal models along with the formal specifications and localizes possible problems. Therefore, it computes all possible trajectories of the entire system behavior. In case of a problem, SESA shows as a result the localization of the error and gives a counterexample. A tool developed in Halle allows for visualization of the error trajectories with Gantt charts.

A further result of this chapter is the verification of a testbed along with safety specifications. Formal models of the testbed have been generated automatically respectively were designed manually. Along with a formal safety specification and the model-checker, the testbed was verified regarding safety (ban of clashing of objects).

New results in this work are the following. In comparison to older works, where the model-form and the approach are similar to this work, one now uses SysML descriptions for gaining a formal specification after all and for the generation of IEC 61499 Function Blocks, which then are used to generate formal models in NCES automatically. One applies the distribution of the controllers and the plant models by using a coordinator control pattern along with the task controller concept. The task controllers are verified without knowledge of the specific coordinator. If the correct behavior of the task controllers is verified, the plant parts together with task controllers are encapsulated in automation objects. This finally results in the usage of libraries, thanks to the modularity of the model-forms and the framework, which in turn allows for easy reconfiguration, thus, of verified components.

6   Integration

As one performed formal verification within the engineering framework and also simulated the system in many stages of the development process, it should not be needed to test the designed system. This is a claim of each framework including formal methods for verification. But to round out the framework and to describe the means to implement the designed controllers on real control devices, this will be done in the example of testing.

After the completion of the three steps of the engineering framework described in the chapters before, or at least after designing controllers as described in the second step of the framework, the integration of real control devices along with executable control code into the real plant is the next and final step. But before the control code on the controller(s) is implemented in closed loop with the real plant, some tests should be performed. Software-In-the-Loop (abbr. SIL) tests already have been successfully performed in Chapter 4.5. That means the control code in closed loop with models is executed directly on the engineering and simulation station and tested for this configuration. Now it is desired to make sure that the construct that was built during the last chapters also works properly with the hardware. That means to perform Hardware-In-the-Loop tests (abbr. HIL). Therefore, there are two possibilities available. The first possibility is to execute the control code on the control device and to control the models in closed loop that are still running on the connected engineering and simulation station. This kind of test can be performed to check how the control code behaves when executed on the control device. Issues to test in that way can be the response time of the controller or even if the controller's hardware interface works properly. Another possibility is to use the control device as a remote [16], and to control the real plant with the engineering and simulation station. In this way it is possible to test first the task controllers together with each mechatronic component, and then the coordinator(s) together with the task controllers.

To do so, the models in the system configuration developed in context with the Model-View-Control design pattern need to be substituted with adequate Service Interface Function Blocks.

These are Function Blocks that allow for accessing services provided with the operating system installed on the control device. This can be, for instance, the access on registers of the I²C Bus on the control device to access the physical in- and outputs of the controller. Also the making available of communication interfaces of the controller belongs to these services.

These Function Blocks need to be developed only once and then can be executed on each component of a controller family. The group of Halle developed Service Interface Function Blocks for Java-based controllers NETMASTER series I and II [20] as well as for WRCACRON controllers [22]. Running testbeds controlled with these controllers can be found in [5]. Other groups developed SIFBs for different controllers, where running testbed implementations can be found in [17], [18] and [19]. So, it is imaginable, that vendors of control devices and automation solutions will provide Service Interface Function Blocks for their devices in future, to assure the integration of IEC 61499 Function Blocks into any automation environment.

Another major issue is the integration of controllers into heterogeneous environments. One cannot expect that the industry will replace the approved PLC environments by another generation of IEC 61499 controllers. Therefore, it is essential to smoothly integrate new control devices into standard PLC environments. The integration of new, IEC 61499 compliant devices into existing IEC 61131 control environments has shown up to be rather not complicated [18]. So, one surely can expect the parallel existence of the two standards within automation solutions in the near future.

The issue that was followed in this work is to equip each mechatronic component with a control chip and to execute the verified task controllers on this chip. The benefits are on hands. Each automation structure could be designed intuitively by engineers and the starting up of the plant could be initiated after a definition and verification of production sequences. This is a scenario that is hardly imaginable applying current standards. This, however, is a courageous glance into the future, but if the IEC 61499 standard has overcome its teething troubles and is accepted by major vendors, this could be a possible future scenario in daily automation engineering.

7   Conclusions

In this work the author proposes an engineering framework which meets the demands of modern automation systems today. The claim of this work is the methodical use of current description means along with new design methodologies. Therefore, a framework is introduced which can be split into three parts that work smoothly together, but also are capable to be adopted each in other contexts and other common engineering frameworks.

The first step is to create a system specification, which is not contradictory and consistent. Therefore, the SysML is chosen as an accepted description means. The work shows a methodical procedure how to create such a specification. It starts from the requirements, which are gathered both in a diagram context but also in textual form, thus being associated with each other in a comprehensible way. The requirements are completed by description means of the SysML and directly linked to corresponding diagrams respectively diagram elements, which fulfill such a requirement. An intermediate step therefore is to create structural descriptions of the automation system, following a component-based architecture by splitting the entire system into modules, which can stand alone and are feasible to be carried out into other specifications created using that modeling guidelines. The behavior descriptions in this specification describe the behavior of the entire system as well as of single components, both considering the pure, uncontrolled plant behavior but also the view of the control level. Qualitative specifications both of the plant as of the controllers make a simulation available, which also validates the fulfillment of requirements. Already in the phase of specification, thanks to the modular approach, a hierarchical control environment is specified. A claim of the author is not only to design a specification which is clear and concise, but the approach of how to create the specification further includes the possibility to automatically generate IEC 61499-compliant applications. Nevertheless the approach can be used in an exclusive context.

The next step contains a yet executable, hardware-independent system specification. Therefore, the IEC 61499 standard is used, which claims to provide easy reconfiguration of automation systems, ease of reusability of components and so on. The work meets the advantages of this standard while following certain modeling guidelines in the directions of modeling, controller development and visualization of hardware components. Some methodological approaches for the development of distributed control strategies are introduced and demonstrated in different contexts of different automation structures. The author proposes pragmatic, at once usable design methodologies for reusable control objects which can end up in easy reconfiguration of entire manufacturing control systems.

After all, it can be shown, that it is not out of reach to integrate the new standard into existing automation systems and so obtaining heterogeneous automation systems, both in hard- and software.

One goal of this work is the ability to simulate the desired automation system in each stage, starting from the specification up to the development of control code that runs on the control device. The achievement of this goal is shown. There is the possibility of application of either Software-in-the-Loop as well as Hardware-in-the-loop methods for validating control code.

A concluding point of this work is the formal verification of the developed automation system. Therefore, a formal model is used, which is rather similar to the IEC 61499 used architecture of Function Blocks and is called Net Condition/ Event systems. By being able to automatically generate formal models of the designed controllers and little adjustments of the models created in the second step of this work, it is possible to formally verify functional safety aspects of an automation system, but it can be applied on other sensor/ actuator systems as well.

Although not a primary goal of this work, the convergence between the three steps of this framework is feasible and shown with prototype tools.

Last not least, the results of this work end up in running, laboratory-scaled plants and demonstrators, already being adopted in today's research projects. This shows the feasibility of the introduced modeling guidelines and approaches in this work.

A next step is to adopt this framework either in the context of industrial-scale automation systems, or even in the context of other sectors like the one of embedded systems.

A remaining issue is to develop an engineering development environment, which combines the possibilities of the framework introduced in this work along with the prototype tools used in this work. The result should be a development environment, which almost fully automatically allows walking along the way of the introduced framework.

References

[1]   V. Vyatkin, H.-M. Hanisch, S. Karras, T. Pfeiffer and V. Dubinin: „Rapid Engineering and Reconfiguration of Automation Objects using Formal Verification", Int. J. Manufacturing Research, Vol. 1, No. 4, pp.382–404, 2006.

[2]   SysML Specification: http://sysml.org/docs/specs/OMGSysML-PAS-07-02-03.pdf, visited in May 2010.

[3]   ARTiSAN Software: http://www.artisansw.com, visited in May 2010.

[4]   Telelogic Rhapsody: http://www.telelogic.com, visited in May 2010.

[5]   Automation Technology Labs of the University of Halle-Wittenberg, http://aut.informatik.uni-halle.de/forschung/testbed, visited in May 2010.

[6]   IEC 61499 – Function Blocks for Industrial-Process Measurement and Control Systems – Part 1: Architecture, International Electrotechnical Commission, Geneva, Switzerland 2005.

[7]   IEC 61131 – International Standard IEC 61131-1, Programmable Controllers – Part 1, Architecture, International Electrotechnical Commission, Geneva, Switzerland 1993.

[8]   IEC 61131 – International Standard IEC 61131-3, Programmable Controllers – Part 3, Programming Languages, International Electrotechnical Comission, Geneva, Switzerland 1993.

[9]   A. Zoitl, R. Smodic, C.K. Sünder and G. Grabmair: "Enhanced Real-Time Execution of Modular Control Software based on IEC 61499", IEEE International Conference on Robotics and Automation (ICRA'06) proceedings pp.327-332, Orlando-Florida, USA 2006.

[10]   S. Panjaitan and G. Frey: "Combination of UML Modeling and the IEC 61499 Function Block Concept for the Development of Distributed Automation Systems" IEEE International Conference on Emerging Technologies and Factory Automation (ETFA 2006) proceedings pp.766-773, Prague, Czech Republic 2006.

[11]   J. L. M. Lastra, A. Lobov, L. Godinho and A. Nunes: "Function Blocks for Industrial Process Measurement and Control Systems: IEC 61499 Introduction and Run-Time Platforms", Institute of Production Engineering, Tampere University of Technology, Tampere, Finland 2004.

[12] V. Vyatkin and H.-M. Hanisch: "A Modeling Approach for Verification of IEC 61499 Function Blocks using Net Condition/Event Systems", IEEE Conference on Emerging Technologies and Factory Automation (ETFA'99) proceedings pp.261-269, Barcelona, Spain 1999.

[13] V. Vyatkin: "IEC 61499 Function Blocks for Embedded and Distributed Control Systems Design", O³NEIDA – Instrumentation Society of America, USA 2007.

[14] H.-M. Hanisch and A. Lüder: "A Signal Extension for Petri Nets and its Use in Controller Design", Fundamenta Informaticae No.4, pp.415-431, March 2000.

[15] M. Hirsch, C. Gerber, V. Vyatkin and H.-M. Hanisch: "Design and Implementation of Heterogeneous Distributed Controllers according to the IEC 61499 Standard – A Case Study", IEEE International Conference on Industrial Informatics (INDIN'07) proceedings pp.829-834, Vienna, Austria 2007.

[16] M. Hirsch, V. Vyatkin and H.-M. Hanisch: "IEC 61499 Function Blocks for Distributed Networked Embedded Applications", IEEE International Conference on Industrial Informatics (INDIN'06) proceedings pp.670-675, Singapore 2006.

[17] EnAS Project Demonstrator: http://aut.informatik.uni-halle.de/forschung/enas_demo/, visited in May 2010.

[18] InfoMechatronics and Industrial Automation lab, University of Auckland: http://www.ece.auckland.ac.nz/~vyatkin/mitra_lab.html, visited in May 2010.

[19] Odo Struger Laboratory, Vienna Technical University: http://www.acin.tu-wien.ac.at/forschung/Projekte/255, visited in May 2010.

[20] Elsist S.r.l. Netmaster Controllers: http://www.elsist.it, visited in May 2010.

[21] Siemens Automation and Drives: http://www.automation.siemens.com, visited in May 2010.

[22] Western Reserve Controls: http://www.wrcakron.com/holocon.html, visited in May 2010.

[23] UML Specification, Version 2.2, Object Management Group: http://www.omg.org/technology/documents/formal/uml.htm, visited in May 2010.

[24] M. Hirsch und H.-M. Hanisch: „Systemspezifikation mit SysML für eine Fertigungstechnische Laboranlage", Fachtagung zum Entwurf komplexer Automatisierungssysteme (EKA'08) proceedings pp.23-34, Magdeburg, Germany 2008.

[25] C. Tranoris and K. Thramboulidis: "Integrating UML and the Function Block Concept for the Development of Distributed Applications", IEEE Conference on

Emerging Technologies and Factory Automation (ETFA'03) proceedings pp.277-284, Lisbon, Portugal 2003.

[26] V. Dubinin, V. Vyatkin and T. Pfeiffer: "Engineering of Validatable Automation Systems using UML-FB", Higher Education Institute Letters, No.2, pp.136-146, ISSN 1728-628X, 2004.

[27] S. Panjaitan and G. Frey: "Development Process for Distributed Automation Systems Combining UML and IEC 61499", International Journal of Manufacturing Research, Vol. 2, No. 1, pp.1–20, Inderscience Publishers, 2007.

[28] T. Weilkiens: "Systems Engineering mit SysML/UML", ISBN 3-89864-409-X, dpunkt.verlag, Heidelberg, Germany 2006.

[29] D. Harel: "Statecharts: A Visual Formalism for Complex Systems", Science of Computer Programming, 8(3):231–274, June 1987.

[30] CORFU ESS 1.0: http://seg.ee.upatras.gr/corfu/dev/download.htm, visited in May 2010.

[31] ISaGRAF: http://www.isagraf.com, visited in May 2010.

[32] O³NEIDA Workbench: www.oooneida.org, visited in May 2010.

[33] Function Block Development Kit: http://www.holobloc.com, visited in May 2010.

[34] 4DIAC: http://fordiac.org, visited in May 2010.

[35] G. Black and V. Vyatkin: "Intelligent Component - based Automation of Baggage Handling Systems with IEC 61499", IEEE Transactions on Automation Science, accepted for future publication 2008, ISSN 1545-5955, to appear 2010.

[36] V. Vyatkin, M. Hirsch and H.-M. Hanisch: "Systematic Design and Implementation of Distributed Controllers in Industrial Automation" IEEE International Conference on Emerging Technologies and Factory Automation (ETFA'06) proceedings pp.633-640, Prague, Czech Republic 2006.

[37] OSEK: http://www.osek-vdx.org, visited in May 2010.

[38] S. Panjaitan and G. Frey: Functional Design for IEC 61499 Distributed Control Systems using UML Activity diagrams", proceedings of the International Conference on Instrumentation Communications and Information Technology ICICI 2005 pp.64-70, Bandung, Indonesia, 2005.

[39] D. Missal and H.-M. Hanisch: "Modular Plant Modeling for Distributed Control", IEEE Conference on Systems, Man and Cybernetics (SMC'07) proceedings pp.3475-3480, Montreal, Canada 2007.

[40]   I. Ivanova-Vasileva, C. Gerber and H.-M. Hanisch: "Transformation of IEC 61499 Control Systems to Formal Models", International Conference Automatics and Informatics (CAI'07) proceedings pp.V-5-V-10, Sofia, Bulgaria 2007.

[41]   I. Ivanova-Vasileva, C. Gerber and H.-M. Hanisch: "Basics of Modeling IEC 61499 Function Blocks with Integer-Valued Data Types", 9th IFAC Workshop on Intelligent Manufacturing Systems, proceedings pp.233-238 Szczecin, Poland 2008.

[42]   I. Ivanova-Vasileva, C. Gerber and H.-M. Hanisch: "A Data Processing Model of IEC 61499 Function Blocks with Integer-Valued Data Types" 9th IFAC Workshop on Intelligent Manufacturing Systems, proceedings pp. 239-244, Szczecin, Poland 2008.

[43]   D. Missal, M. Hirsch and H.-M. Hanisch: "Hierarchical Distributed Controllers – Design and Verification", IEEE International Conference on Emerging Technologies and Factory Automation (ETFA'07) proceedings pp.657-664, Patras, Greece 2007.

[44]   P.H. Starke and S. Roch: "Analyzing Signal-Net Systems", Technical Report, No.162 in Informatik-Berichte, Humboldt-Universität zu Berlin, Berlin, Germany 2002.

[45]   V. Vyatkin and H.-M. Hanisch: „Bringing the Model-based Verification of Distributed Control Systems to the Engineering Practice", in Intelligent Manufacturing Systems 2001, M. Zaremba, J. Szpytko, Z. Banaszak (eds.), Poznan, Poland, Elsevier Science, pp.152-157, 2001.

[46]   M. Hirsch, D. Missal and H.-M. Hanisch: "Design and Verification of Distributed Industrial Manufacturing Control Systems", Annual Conference of the IEEE Industrial Electronics Society (IECON'08) proceedings pp.152-157, Orlando-Florida, USA 2008.

[47]   J.H. Christensen: "Design Patterns for Systems Engineering with IEC 61499", Verteilte Automatisierung, proceedings, Otto-von-Guericke-University Magdeburg, Germany 2000.

[48]   FlexRay specification: http://www.flexray.com, visited in May 2010.

[49]   S. Preuße and H.-M. Hanisch: "Specification and Verification of Technical Plant Behavior with Symbolic Timing Diagrams", International Design and Test Workshop (ITD) proceedings pp.313-318, Monastir, Tunisia 2008.

[50]   X. Cai, V. Vyatkin and H.-M. Hanisch: "Design and Implementation of a Prototype Control System According to IEC 61499", Conference on Emerging Technologies and Factory Automation (ETFA'03) proceedings pp.269-276, Lisbon, Portugal 2003.

[51]   C.C. Insaurralde, M.A. Seminario, J.F. Jimenez and J.M. Giron-Sierra: "IEC 61499 Model for Avionic Distributed Fuel Systems with Networked Embedded Holonic Controllers", Conference on Emerging Technologies and Factory Automation (ETFA'06) proceedings pp.388-396, Prague, Czech Republic 2006.

[52]   S. Panjaitan: "Development Process for Distributed Automation Systems based on Elementary Mechatronic Functions", PhD Thesis, ISBN 978-3-8322-6934-0, Shaker Publishing, Aachen, Germany 2008.

[53]   V. Vyatkin, Z. Salcic, P.S. Roop and J. Fitzgerald: "Now that's smart", IEEE Industrial Electronics Magazine, 1 (4), pp.17-29, 2007.

[54]   T. Baier, J. Fritsche, G. Keintzel, D. Loy, R. Schranz, H. Steininger, T. Strasser and C. Sünder: "Future Scenarios for Application of Downtimeless Reconfiguration in Industrial Practice", International Conference in Industrial Informatics (INDIN'07) proceedings pp.1129-1134, Vienna, Austria 2007.

[55]   M. Colla, A. Brusaferri and E. Carpanzano: "Applying the IEC 61499 Model to the Shoe Manufacturing Sector", Conference on Emerging Technologies and Factory Automation (ETFA'06) proceedings pp.1301-1308, Prague, Czech Republic 2006.

[56]   C. Gerber, M. Hirsch and H.-M. Hanisch: "Automatisierung einer energieautarken Fertigungsanlage nach IEC 61499", Automatisierungstechnische Praxis (atp 03/09) pp.44-52, Munich, Germany 2009.

[57]   V. Dubinin and V. Vyatkin: "On Definition of a Formal Semantic Model for IEC 61499 Function Blocks", EURASIP Journal of Embedded Systems, vol. 2008, Article ID 426713, 10 pp., 2008. doi:10.1155/2008/426713.

[58]   V. Vyatkin, V. Dubinin, L.M. Ferrarini and C. Veber: "Alternatives for Execution Semantics of IEC61499", IEEE Conference on Industrial Informatics (INDIN'2007), Vienna, Austria 2007.

[59]   G. Frey and T. Hussain: "Modeling Techniques for Distributed Control Systems based on the IEC 61499 Standard - Current Approaches and Open Problems", International Workshop on Discrete-event Systems (WODES 2006) proceedings pp. 176-181, Ann Arbor, Michigan, USA 2006.

[60] C.K. Sünder, A. Zoitl, J. Christensen, M. Colla and T. Strasser: "Execution Models for the IEC 61499 Elements Composite Function Block and Subapplication", IEEE International Conference on Industrial Informatics (INDIN'07) proceedings pp.1169-1175, Vienna, Austria 2007.

[61] O³Neida work group on compliance profile: http://www.oooneida.org/standards_development_Compliance_Profile.html, visited in May 2010.

[62] M. Heiner, T. Mertke and P. Deussen: "A Safety-oriented Technical Language for the Requirement Specification in Control Engineering", Computer Science Reports 09/01, page 65 ff, BTU Cottbus, Mai 2001.

[63] S. Preuße and H.-M. Hanisch: "Specification of Technical Plant Behavior with a Safety-Oriented Technical Language", 7th IEEE International Conference on Industrial Informatics (INDIN'09) proceedings pp.632-637, Cardiff, United Kingdom 2009.

[64] S. Roch: "Extended Computation Tree Logic", CS&P workshop, Informatik-Berichte no.140, proceedings pp.225-234, Berlin, Germany 2004.

[65] A. Fehnker, R. Huuck, B. Schlich and M. Tapp: "Automatic Bug Detection in Microcontroller Software by Static Program Analysis" 35th International Conference on Current Trends in Theory and Practice of Computer Science (SOFSEM) proceedings pp.267-278, Czech Republic 2009.

[66] S. Bornot, R. Huuck, Y. Lakhnech and B. Lukoschus: "Utilizing Static Analysis for Programmable Logic Controllers", The 4th International Conference on Automation of Mixed Processes: Hybrid Dynamic Systems proceedings pp.183-187, Dortmund, Germany 2000.

[67] Method for Performing Verification of Logic Circuits, United States Patent 7398494, http://www.freepatentsonline.com/7398494.html, visited in May 2010.

[68] H.-M. Hanisch: "Closed-Loop Modeling and Related Problems of Embedded Control Systems in Engineering", Lecture Notes in Computer Science, Vol. 3052, pp.6-19, Springer, 2004.

[69] T. Heverhagen, R. Tracht and R. Hirschfeld: "A Profile for Integrating Function Blocks into the Unified Modeling Language", International Workshop SVERTS'03, Proceedings, San Francisco, California 2003.

[70]   Tanvir Hussain: "Development and Automatic Deployment of Distributed Control Applications", PhD Thesis, Shaker Verlag, Aachen 2009, ISBN: 978-3-8322-7969-1.

[71]   C. Gerber, I. Ivanova-Vasileva and H.-M. Hanisch: "Formal Modeling of IEC 61499 Function Blocks with Integer-valued Data Types", Control and Cybernetics, 39(2010), No. 1, pp. 197 - 231, Warsaw, Poland 2010.

Biographical Sketch

Martin Hirsch
Born in 1980 in Halle (Saale)

Education

1986-1991	Poly-technical High school
1991-1998	Secondary High school
1998	General Qualification for University Entrance (Abitur)

1998-1999	Military Service
	German Federal Armed Forces
	Air Force Infantry

Study

2000-2006	Study of Engineering Sciences/ Engineering Informatics
	Martin Luther University of Halle-Wittenberg
2004-2005	Student Research Assistant
	Chair for Automation Technology
	Martin Luther University of Halle-Wittenberg
2005-2006	Honorary Research Fellow
	Department of Electrical and Computer Engineering
	University of Auckland, New Zealand
2006	Degree: Diplom-Ingenieur, Receipt of Master's Equivalent
2010	Defense of Doctoral Thesis, Degree: Doktor-Ingenieur

Professional Life

2006-2007	Scientific Staff
	Department of Engineering Sciences
	Martin Luther University of Halle-Wittenberg
2007-2011	Scientific Staff
	Institute for Computer Science
	Martin Luther University of Halle-Wittenberg

Abstract

Currently, there are numerous challenges in the Automation Technology sector regarding such key words like distribution of controllers and easy reconfiguration of systems. That is namely flexibility, which needs easy portability of software to different types of hardware and, linked therewith, component-based encapsulation of Intellectual Property.

Also other engineering areas, e.g. the traditional embedded systems domain, often suffer from malfunctions due to increasing amounts of electrical devices and their distribution and therefore, increasing complexity. Malfunctions in these areas mostly occur because of suboptimal requirements engineering, including the usage of inconsistent specifications, and no verification of the control software. A conclusion is to optimize the Requirements engineering/ Specifications along with adequate means and to apply formal verification methods along with well-fitting system description opportunities. One issue of this work is to cover the wish of engineers to adopt parts of the specification for the design of the controllers. Another aspect is to use means for generic higher-level system description, which supports such mentioned features like distribution, reconfiguration, reusability and interoperability. Naturally, an increasing degree of distribution also implies an increasing complexity of the entire system. To master this complexity, adequate systematic methods need to be developed and applied. In this work, a main focus is a way to design highly reusable controllers.

The chapter on verification is a sketch since it is not the main focus of the author, but it is accomplished by other members of the work group in Halle. The goal of this work is to provide a methodological framework to integrate formal methods into an engineering framework focusing on systematic design of distributed controllers following the IEC 61499 international standard. The results of the application of the introduced framework are demonstrated on an exemplary testbed.

The final result of this work is a systematic approach to design and implement distributed controllers, from scratch up to integration, covering the important issues of distribution, reusability, flexibility and reliability.

Kurzfassung

In der Automatisierungstechnik tauchen in den letzten Jahren vermehrt Schlüsselwörter wie verteilte Automatisierung, einfache Rekonfigurierbarkeit sowie Wiederverwendbarkeit von Automatisierungskomponenten auf. Zusammengefasst bedeutet das große neue Herausforderungen an die Hersteller von Automatisierungslösungen, um die neu geforderte Flexibilität gewährleisten zu können. So müssen zum Einen Spezifikationstechniken entwickelt und angewendet werden, die der Komplexität verteilter Systeme gerecht werden und nicht schon in der Phase der Spezifizierung für erhöhte Verwirrung sorgen. Zum Anderen müssen aber auch aktuelle Lösungen für die softwaretechnische Realisierung der Automatisierung nach und nach durch neuere, flexiblere Ansätze ersetzt werden. Die Möglichkeit der Verifizierbarkeit von Automatisierungslösungen sollte in diesem Kontext auch nicht außer Acht gelassen werden.

In den letzten Jahren haben sich verschiedene Technologien zum modellbasierten Steuerungsentwurf herausgebildet, auf welche in dieser Arbeit eingegangen wird. Diese Technologien werden methodisch miteinander verknüpft, um eine glatte Ingenieursarbeit ohne Bruchstellen zu gewährleisten. Die einzelnen Schritte der hier vorgestellten Vorgehensweise können aber auch entkoppelt voneinander betrachtet und in einem eigenen Kontext bewertet werden.

In dieser Arbeit wird auf die Spezifizierung von Systemen mit der Sprache SysML eingegangen, welche bei sinnvoller und ausgesuchter Anwendung etliche Vorteile mit sich bringt. So werden textuell formulierte Anforderungen systematisch mit Diagrammen unterschiedlicher Diagrammformen beschrieben, welche sowohl die Struktur als auch das Verhalten eines Systems beschreiben können. Hierbei werden durch Einhaltung von Modellierungsrichtlinien Eindeutigkeit und Konsistenz gewährleistet.

Die softwaretechnische Lösung zur Automatisierung wird in dieser Arbeit mit dem Standard 61499 der IEC vorgenommen. Dieser Standard ist eine streng komponentenorientierte Weiterentwicklung der IEC 61131 und wurde speziell zum Entwurf verteilter Systeme entwickelt. Die zuvor erstellte Spezifikation kann hierbei entweder komplett verwendet werden, um Systemkonfigurationen zu generieren, es ist aber auch möglich, nur Teile der Spezifikation zum Systementwurf zu verwenden. Der Standard bietet die Möglichkeit ein System funktional vollständig zu beschreiben, Applikationen können frei im System verteilt und integriert werden. Durch die Portierbarkeit der ereignisgetriebenen

Steuerungsapplikationen auf verschiedene Hardware ergibt sich die Möglichkeit der Interoperabilität von Geräten verschiedener Hersteller. In dieser Arbeit wird speziell auf Vorgehensweisen zum verteilten Steuerungsentwurf eingegangen, die die Wiederverwendbarkeit von Steuerungskomponenten im Besonderen gewährleisten.

Die Möglichkeit der Verifizierbarkeit des modellbasierten Steuerungsentwurfes ergibt sich unter Verwendung von Net-Condition/Event Systemen (NCES). Diese Modellform mit formalem Hintergrund beinhaltet neben der Eigenschaft der Modularität auch Ereignis- und Datenverbindungen und fügt sich somit nahtlos der IEC 61499 an.

Die nach IEC 61499 erstellten Modelle der Strecke werden im letzten Schritt mit Service Interface Function Blocks substituiert. Diese ermöglichen den Zugriff auf die verwendete Steuerungshardware.